Edited by Jack Dann

UNICORNS!
MAGICATS!
BESTIARY!
MERMAIDS!
SORCERERS!
DEMONS!
DOGTALES!
SEASERPENTS!
DINOSAURS!
LITTLE PEOPLE!
MAGICATS II
UNICORNS II
DRAGONS!
INVADERS!
HORSES!
ANGELS!
HACKERS
TIMEGATES
CLONES
IMMORTALS
NANOTECH
FUTURE WAR
SPACE SOLDIERS
FUTURE SPORTS
BEYOND FLESH

BEYOND FLESH

EDITED BY
JACK DANN & GARDNER DOZOIS

ACE BOOKS, NEW YORK

BEYOND FLESH

An Ace Book / published by arrangement with the editors

PRINTING HISTORY
Ace mass-market edition / December 2002

Copyright © 2002 by Jack Dann and Gardner Dozois.
Cover art by Jan Franz/Gettyimages.
Cover design by Rita Frangie.

Visit our website at
www.penguinputnam.com
Check out the ACE Science Fiction & Fantasy newsletter!

ISBN: 0-441-00999-9

ACE®
Ace Books are published by The Berkley Publishing Group,
a division of Penguin Putnam Inc.,
375 Hudson Street, New York, New York 10014.
ACE and the "A" design
are trademarks belonging to Penguin Putnam Inc.

PRINTED IN THE UNITED STATES OF AMERICA

10 9 8 7 6 5 4 3 2 1

Acknowledgment is made for permission to reprint the following material:

"Call Me Joe," by Poul Anderson. Copyright © 1957 by Street & Smith Publications, Inc. First published in *Astounding*. Reprinted by permission of the author's estate.

"Learning to Be Me," by Greg Egan. Copyright © 1990 by Interzone. First published in *Interzone 37*, July 1990. Reprinted by permission of the author.

"Pretty Boy Crossover," by Pat Cadigan. Copyright © 1986 by Davis Publications, Inc. First published in *Asimov's Science Fiction*, January 1986. Reprinted by permission of the author.

"Ancient Engines," by Michael Swanwick. Copyright © 1998 by Dell Magazines. First published in *Asimov's Science Fiction*, February 1999. Reprinted by permission of the author.

"Winemaster," by Robert Reed. Copyright © 1999 by Mercury Press, Inc. First published in *The Magazine of Fantasy & Science Fiction*, July 1999. Reprinted by permission of the author.

"More Adventures on Other Planets," by Michael Cassutt. Copyright © 2001 by SCIFI.COM. First published electronically on SCI FICTION, January 10, 2001. Reprinted by permission of the author.

"Nevermore," by Ian R. MacLeod. Copyright © 1997 by Ian R. MacLeod. First published in *Dying For It* (HarperPrism). Reprinted by permission of the author.

"Approaching Perimelasma," by Geoffrey A. Landis. Copyright © 1998 by Dell Magazines. First published in *Asimov's Science Fiction*, January 1998. Reprinted by permission of the author.

"The Gravity Mine," by Stephen Baxter. Copyright © 2000 by Dell Magazines. First published in *Asimov's Science Fiction*, April 2000. Reprinted by permission of the author.

"Reef," by Paul J. McAuley. Copyright © 2000 by Paul J. McAuley. First published in *Skylife* (Harcourt), edited by Gregory Benford and George Zebrowski. Reprinted by permission of the author.

CONTENTS

PREFACE

Ever since human beings evolved enough for us to become aware of ourselves as sentient beings with a physical existence in the universe, to know ourselves as living creatures alive and abroad in the world, we have *also* been made aware of the *limitations* of that physical existence. Lessons in the limitations of the flesh have been painfully ground home over so many countless millennia that we long ago came to take them for granted: You can't pick a hot coal out of the fire without being burned, you can't jump off a cliff and fly, you can't breathe underwater, you can't exist for long without food or water or sleep, there are weights too heavy to lift and distances too broad to leap, you can't go where it's too cold without freezing to death, or where it's too hot without passing out and dying. You can't see what's over the next hill or in the next valley without *going* there to look. You can't make yourself heard farther away than your voice will reach. You can't be in two places at once. You must grow old. You must die.

These limitations have always *fretted* human beings, though, and ever since the earliest beginnings of human culture, clever men and women have been trying to think up ways to get *around* them. Technology itself, in its most basic forms, can be seen as methods to get around the fundamental limitations of the flesh: fire and clothing made from animal furs to allow us to venture into the deepest winter (or even Ice Age) climates and survive, levers and pulleys to help us manipulate weights too heavy to be lifted, tools and knives to help us rip flesh and cut wood and stone when the teeth and claws that Nature started us out with as part of our basic human tool-kit proved inadequate to the jobs we wanted to *do* with them.

As technology increased in sophistication and power, so we have steadily been able to transcend more and more of the limitations of the flesh, of the basic human tool-kit, the basic human form, that we were all issued at birth—so that now, here at the beginning of the twenty-first century, you *can* jump off a cliff and fly (if you have the right technology, anyway!), and you can *also* fly higher and longer and faster than any bird possibly could, fly right out of the atmosphere, in fact, and into outer space; you can exist underwater for hours, and may someday be able to live underwater for as long as you'd like; you can lift and manipulate and ship over long distances weights so immense and unwieldy that even thousands of humans all straining together couldn't shift them; you can talk to people and make your voice clearly heard even if they're thousands of miles away, you can listen to the voices of the dead, and record your *own* voice so that it can be listened to someday by people who have yet to be born; and, without leaving your chair, you *can* see what's happening over the next hill or in the next valley, or on the other side of the world, or in the depths of interstellar space, or even in the secret and heretofore inaccessible inner regions of your own body.

And you ain't seen *nothing* yet!

As the anthology that follows demonstrates, here at the beginning of the twenty-first century we stand poised on the brink of scientific revolutions that may do away with all the *rest* of the limitations of the flesh as well, even fundamental and seemingly unchallengeable ones such as not being able to go where flesh would burn or wither or freeze, or where there is no air for our lungs to breathe (or where the "air" is poison), or where the gravity is strong enough to crush an ordinary human like a bug. Or needing food and water or sleep. Or having to age. Or having to *die*. Or even not being able to be in two places at once!

So open the pages of this book, and let some of the world's most expert dreamers show you what it might be like when human consciousness is no longer restricted to the prison of the flesh, or of the basic human form . . .

where you can sit safely at home and still explore the
deadly frozen ice-fields of Europa or the swirling poison-
gas hell of Jupiter . . . where you can download your consciousness into a computer, or into an artificial
body-environment the size of a molecule . . . or into a
robot body that could last forever . . . or replicate yourself
in a hundred different forms in a hundred different envi-
ronments . . . or find a love that persists, and is recipro-
cated, even after death . . . or leave not only the flesh but
all physical matter behind, and roam through space as dis-
corporate intelligences until the end of the universe itself . . .

Enjoy!

(For more speculations on these themes, check out our
Ace anthologies: *Genometry, Immortals, Clones, Nanotech, Hackers, Future War,* and *Space Soldiers.*)

CALL ME JOE

Poul Anderson

Here's one of the earliest explorations in science fiction of the idea that the human spirit, if indomitable enough, may find ways to move beyond flesh and transcend the limitations of the human form to go where a frail mortal body cannot go—and, forty-five years after its first publication, still one of the best.

One of the best-known writers in science fiction, the late Poul Anderson made his first sale in 1947, while he was still in college, and in the course of his subsequent fifty-four-year career published almost a hundred books (in several different fields, as Anderson wrote historical novels, fantasies, and mysteries, in addition to SF), sold hundreds of short pieces to every conceivable market, and won seven Hugo Awards, three Nebula Awards, and the Tolkein Memorial Award for life achievement.

Anderson had trained to be a scientist, taking a degree in physics from the University of Minnesota, but the writing life proved to be more seductive, and he never did get around to working in his original field of choice. Instead, the sales mounted steadily, until by the late '50s and early '60s he may have been one of the most prolific writers in the genre.

In spite of his high output of fiction, he somehow managed to maintain an amazingly high standard of literary quality as well, and by the early to mid '60s he was also on his way to becoming one of the most honored and respected writers in the genre. At one point during this period (in addition to non-related work, and lesser series such as the "Hoka" stories he was writing in collaboration with Gordon R. Dickson), Anderson was running three of the most popular and prestigious series in science fiction all at the same time: *the "Technic History" series detailing the exploits of the wily trader Nicholas Van Rijn (which includes novels such as* The Man Who Counts, The Trouble Twisters, Satan's World, Mirkheim, The People of the Wind, *and collections such as* Trader to the Stars *and* The Earth Book of Stormgate*); the extremely popular series relating the adventures of interstellar secret agent Dominic Flandry, probably the most successful attempt to cross SF with the spy thriller, next to Jack Vance's "Demon Princes" novels (the Flandry series includes novels such as* A Circus of Hells, The Rebel Worlds, The Day of Their Return, Flandry of Terra, A Knight of Ghosts and Shadows, A Stone in Heaven, *and* The Game of Empire, *and collections such as* Agent of the Terran Empire*); and, my own personal favorite, a series that took us along on assignment with the agents of the Time Patrol (including the collections* The Guardians of Time, Time Patrolman, The Shield of Time, *and* The Time Patrol*).*

When you add to this amazing collection of memorable titles the impact of the best of Anderson's non-series novels, work such as Brain Wave, Three Hearts and Three Lions, The Night Face, The Enemy Stars, *and* The High Crusade, *all of which was being published in* addition *to the series books, it becomes clear that Anderson dominated the late '50s and the pre-New Wave '60s in a way that only Robert A. Heinlein, Isaac Asimov, and Arthur C. Clarke could rival. Anderson, in fact, would continue to be an active and dominant figure for the rest of the twentieth century and on into the next, continuing to produce strong and innovative work until the very end of his life, winning the John W. Campbell Award for his novel* Genesis *just months before his death.*

Anderson's other books (among many *others) include:* The Broken Sword, Tau Zero, A Midsummer Tempest, Orion Shall Rise, The Boat of a Million Years, Harvest of

Stars, The Fleet of Stars, Starfarers, *and* Operation Luna. *His short work has been collected in* The Queen of Air and Darkness and Other Stories, Fantasy, The Unicorn Trade *(with Karen Anderson),* Past Times, The Best of Poul Anderson, Explorations, *and, most recently, the retrospective collection* All One Universe. *The last book published in his lifetime was a new novel,* Genesis; *there are several other new books in the pipeline that will appear posthumously. He died in 2001.*

T*he wind came* whooping out of eastern darkness, driving a lash of ammonia dust before it. In minutes, Edward Anglesey was blinded.

He clawed all four feet into the broken shards which were soil, hunched down, and groped for his little smelter. The wind was an idiot bassoon in his skull. Something whipped across his back, drawing blood, a tree yanked up by the roots and spat a hundred miles. Lightning cracked, immensely far overhead where clouds boiled with night.

As if to reply, thunder toned in the ice mountains and a red gout of flame jumped and a hillside came booming down, spilling itself across the valley. The earth shivered.

Sodium explosion, thought Anglesey in the drumbeat noise. The fire and the lightning gave him enough illumination to find his apparatus. He picked up tools in muscular hands, his tail gripped the trough, and he battered his way to the tunnel and thus to his dugout.

It had walls and roof of water, frozen by sun-remoteness and compressed by tons of atmosphere jammed onto every square inch. Ventilated by a tiny smokehole, a lamp of tree oil burning in hydrogen made a dull light for the single room.

Anglesey sprawled his slate-blue form on the floor, panting. It was no use to swear at the storm. These ammonia gales often came at sunset, and there was nothing to do but wait them out. He was tired anyway.

It would be morning in five hours or so. He had hoped to cast an axehead, his first, this evening, but maybe it was better to do the job by daylight.

He pulled a decapod body off a shelf and ate the meat raw, pausing for long gulps of liquid methane from a jug. Things would improve once he had proper tools; so far, everything had been painfully grubbed and hacked to shape with teeth, claws, chance icicles, and what detestably weak and crumbling fragments remained of the spaceship. Give him a few years and he'd be living as a man should.

He sighed, stretched, and lay down to sleep.

Somewhat more than one hundred and twelve thousand miles away, Edward Anglesey took off his helmet.

He looked around, blinking. After the Jovian surface, it was always a little unreal to find himself here again, in the clean, quiet orderliness of the control room.

His muscles ached. They shouldn't. He had not really been fighting a gale of several hundred miles an hour, under three gravities, and a temperature of 140 Absolute. He had been here, in the almost nonexistent pull of Jupiter V, breathing oxynitrogen. It was Joe who lived down there and filled his lungs with hydrogen and helium at a pressure which could still only be estimated because it broke aneroids and deranged piezo-electrics.

Nevertheless, his body felt worn and beaten. Tension, no doubt—psychosomatics—after all, for a good many hours now he had, in a sense, been Joe, and Joe had been working hard.

With the helmet off, Anglesey held only a thread of identification. The esprojector was still tuned to Joe's brain but no longer focused on his own. Somewhere in the back of his mind, he knew an indescribable feeling of sleep. Now and then, vague forms or colors drifted in the soft black—dreams? Not impossible, that Joe's brain should dream a little when Anglesey's mind wasn't using it.

A light flickered red on the esprojector panel, and a bell whined electronic fear. Anglesey cursed. Thin fingers danced over the controls of his chair, he slued around and shot across to the bank of dials. Yes—there—K-tube oscillating again! The circuit blew out. He wrenched the faceplate off with one hand and fumbled in a drawer with the other.

Inside his mind he could feel the contact with Joe fading. If he once lost it entirely, he wasn't sure he could regain it. And Joe was an investment of several million dollars and quite a few highly skilled man-years.

Anglesey pulled the offending K-tube from its socket and threw it on the floor. Glass exploded. It eased his temper a bit, just enough so he could find a replacement, plug it in, switch on the current again—as the machine warmed up, once again amplifying, the Joeness in the back alleys of his brain strengthened.

Slowly, then, the man in the electric wheelchair rolled out of the room, into the hall. Let somebody else sweep up the broken tube. To hell with it. To hell with everybody.

Jan Cornelius had never been farther from Earth than some comfortable Lunar resort. He felt much put upon that the Psionics Corporation should tap him for a thirteen-month exile. The fact that he knew as much about esprojectors and their cranky innards as any other man alive was no excuse. Why send anyone at all? Who cared?

Obviously the Federation Science Authority did. It had seemingly given those bearded hermits a blank check on the taxpayers' account.

Thus did Cornelius grumble to himself, all the long hyperbolic path to Jupiter. Then the shifting accelerations of

approach to its tiny inner satellite left him too wretched for
further complaint.

And when he finally, just prior to disembarkation, went
up to the greenhouse for a look at Jupiter, he said not a
word. Nobody does, the first time.

Arne Viken waited patiently while Cornelius stared. *It
still gets me, too,* he remembered. *By the throat. Sometimes
I'm afraid to look.*

At length Cornelius turned around. He had a faintly
Jovian appearance himself, being a large man with an im-
posing girth. "I had no idea," he whispered. "I never
thought . . . I had seen pictures, but—"

Viken nodded. "Sure, Dr. Cornelius. Pictures don't convey it."

Where they stood, they could see the dark broken rock
of the satellite, jumbled for a short way beyond the landing slip and then chopped off sheer. This moon was
scarcely even a platform, it seemed, and cold constella-
tions went streaming past it, around it. Jupiter lay across a
fifth of that sky, softly ambrous, banded with colors, spotted with the shadows of planet-sized moons and with
whirlwinds as broad as Earth. If there had been any gravity to speak of, Cornelius would have thought, instinctively, that the great planet was falling on him. As it was,
he felt as if sucked upward; his hands were still sore where
he had grabbed a rail to hold on.

"You live here . . . all alone . . . with this?" He spoke
feebly.

"Oh, well, there are some fifty of us all told, pretty con-
genial," said Viken. "It's not so bad. You sign up for four-
cycle hitches—four ship arrivals— and believe it or not,
Dr. Cornelius, this is my third enlistment."

The newcomer forbore to inquire more deeply. There
was something not quite understandable about the men on
Jupiter V. They were mostly bearded, though otherwise
careful to remain neat; their low-gravity movements were

somehow dreamlike to watch; they hoarded their conversation, as if to stretch it through the year and month between ships. Their monkish existence had changed them—or did they take what amounted to vows of poverty, chastity, and obedience because they had never felt quite at home on green Earth?

Thirteen months! Cornelius shuddered. It was going to be a long, cold wait, and the pay and bonuses accumulating for him were scant comfort now, four hundred and eighty million miles from the sun.

"Wonderful place to do research," continued Viken. "All the facilities, handpicked colleagues, no distractions . . . and of course—" He jerked his thumb at the planet and turned to leave.

Cornelius followed, wallowing awkwardly. "It is very interesting, no doubt," he puffed. "Fascinating. But really, Dr. Viken, to drag me way out here and make me spend a year plus waiting for the next ship . . . to do a job which may take me a few weeks—"

"Are you sure it's that simple?" asked Viken gently. His face swiveled around, and there was something in his eyes that silenced Cornelius. "After all my time here, I've yet to see any problem, however complicated, which when you looked at it the right way didn't become still more complicated."

They went through the ship's air lock and the tube joining it to the station entrance. Nearly everything was underground. Rooms, laboratories, even halls had a degree of luxuriousness—why, there was a fireplace with a real fire in the common room! God alone knew what *that* cost!

Thinking of the huge chill emptiness where the king planet laired, and of his own year's sentence, Cornelius decided that such luxuries were, in truth, biological necessities.

Viken showed him to a pleasantly furnished chamber which would be his own. "We'll fetch your luggage soon and unload your psionic stuff. Right now, everybody's either talking to the ship's crew or reading his mail."

Cornelius nodded absently and sat down. The chair, like all low-gee furniture, was a mere spidery skeleton, but it held his bulk comfortably enough. He felt in his tunic hoping to bribe the other man into keeping him company for a while. "Cigar? I brought some from Amsterdam."

"Thanks." Viken accepted with disappointing casualness, crossed long thin legs, and blew grayish clouds.

"Ah . . . are you in charge here?"

"Not exactly. No one is. We do have one administrator, the cook, to handle what little work of that type may come up. Don't forget, this is a research station, first, last, and always."

"What is your field, then?"

Viken frowned. "Don't question anyone else so bluntly, Dr. Cornelius," he warned. "They'd rather spin the gossip out as long as possible with each newcomer. It's a rare treat to have someone whose every last conceivable reaction hasn't been—No, no apologies to me. 'S all right. I'm a physicist, specializing in the solid state at ultrahigh pressures." He nodded at the wall. "Plenty of it to be observed—there!"

"I see." Cornelius smoked quietly for a while. Then: "I'm supposed to be the psionics expert, but frankly, at present, I've no idea why your machine should misbehave as reported."

"You mean those, uh, K-tubes have a stable output on Earth?"

"And on Luna, Mars, Venus . . . everywhere, apparently, but here." Cornelius shrugged. "Of course, psibeams are always persnickety, and sometimes you get an unwanted feedback when—No. I'll get the facts before I theorize. Who are your psimen?"

"Just Anglesey, who's not a formally trained esman at all. But he took it up after he was crippled, and showed such a natural aptitude that he was shipped out here when he volunteered. It's so hard to get anyone for Jupiter V that we aren't fussy about degrees. At that, Ed seems to be operating Joe as well as a Ps.D. could."

"Ah, yes. Your pseudojovian. I'll have to examine that angle pretty carefully too," said Cornelius. In spite of himself, he was getting interested. "Maybe the trouble comes from something in Joe's biochemistry. Who knows? I'll let you into a carefully guarded little secret, Dr. Viken: Psionics is not an exact science."

"Neither is physics," grinned the other man. After a moment, he added more soberly: "Not my brand of physics, anyway. I hope to make it exact. That's why I'm here, you know. It's the reason we're all here."

Edward Anglesey was a bit of a shock, the first time. He was a head, a pair of arms, and a disconcertingly intense blue stare. The rest of him was mere detail, enclosed in a wheeled machine.

"Biophysicist originally," Viken had told Cornelius. "Studying atmospheric spores at Earth Station when he was still a young man—accident crushed him up, nothing below his chest will ever work again. Snappish type, you have to go slow with him."

Seated on a wisp of stool in the esprojector control room, Cornelius realized that Viken had been soft-pedaling the truth.

Anglesey ate as he talked, gracelessly, letting the chair's tentacles wipe up after him. "Got to," he explained. "This stupid place is officially on Earth time, GMT. Jupiter isn't. I've got to be here whenever Joe wakes, ready to take him over."

"Couldn't you have someone spell you?" asked Cornelius.

"Bah!" Anglesey stabbed a piece of prot and waggled it

at the other man. Since it was native to him, he could spit
out English, the common language of the station, with un-
measured ferocity. "Look here. You ever done therapeutic
esping? Not just listening in, or even communication, but
actual pedagogic control?"

"No, not I. It requires a certain natural talent, like
yours." Cornelius smiled. His ingratiating little phrase was
swallowed without being noticed by the scored face opposite him. "I take it you mean cases like, oh, reeducating the
nervous system of a palsied child?"

"Yes, yes. Good enough example. Has anyone ever
tried to suppress the child's personality, take him over in
the most literal sense?"

"Good God, no!"

"Even as a scientific experiment?" Anglesey grinned.
"Has any esprojector operative ever poured on the juice
and swamped the child's brain with his own thoughts?
Come on, Cornelius, I won't snitch on you."

"Well . . . it's out of my line, you understand." The
psionicist looked carefully away, found a bland meter face,
and screwed his eyes to that. "I have, uh, heard something
about . . . well, yes, there were attempts made in some
pathological cases to, uh, bull through . . . break down the
patient's delusions by sheer force—"

"And it didn't work," said Anglesey. He laughed. "It
can't work, not even on a child, let alone an adult with a
fully developed personality. Why, it took a decade of re-
finement, didn't it, before the machine was debugged to
the point where a psychiatrist could even 'listen in' without the normal variation between his pattern of thought
and the patient's . . . without that variation setting up an interference scrambling the very thing he wanted to study.
The machine has to make automatic compensations for the
differences between individuals. We still can't bridge the
differences between species.

"If someone else is willing to cooperate, you can very

gently guide his thinking. And that's all. If you try to seize control of another brain, a brain with its own background of experience, its own ego—you risk your very sanity. The other brain will fight back, instinctively. A fully developed, matured, hardened human personality is just too complex for outside control. It has too many resources, too much hell the subconscious can call to its defense if its integrity is threatened. Blazes, man, we can't even master our own minds, let alone anyone else's!"

Anglesey's cracked-voice tirade broke off. He sat brooding at the instrument panel, tapping the console of his mechanical mother.

"Well?" said Cornelius after a while.

He should not, perhaps, have spoken. But he found it hard to remain mute. There was too much silence—half a billion miles of it, from here to the sun. If you closed your mouth five minutes at a time, the silence began creeping in like a fog.

"Well," gibed Anglesey. "So our pseudojovian, Joe, has a physically adult brain. The only reason I can control him is that his brain has never been given a chance to develop its own ego. I *am* Joe. From the moment he was 'born' into consciousness, I have been there. The psibeam sends me all his sense data and sends him back my motor-nerve impulses. But nevertheless, he has that excellent brain, and its cells are recording every trace of experience, even as yours and mine; his synapses have assumed the topography which is my 'personality pattern.'

"Anyone else, taking him over from me, would find it was like an attempt to oust me myself from my own brain. It couldn't be done. To be sure, he doubtless has only a rudimentary set of Anglesey memories—I do not, for instance, repeat trigonometric theorems while controlling him—but he has enough to be, potentially, a distinct personality.

"As a matter of fact, whenever he wakes up from sleep—there's usually a lag of a few minutes, while I sense the change through my normal psi faculties and get the amplifying helmet adjusted—I have a bit of a struggle. I feel almost a . . . a resistance . . . until I've brought his mental currents completely into phase with mine. Merely dreaming has been enough of a different experience to—"

Anglesey didn't bother to finish the sentence.

"I see," murmured Cornelius. "Yes, it's clear enough. In fact, it's astonishing that you can have such total contact with a being of such alien metabolism."

"I won't for much longer," said the esman sarcastically, "unless you can correct whatever is burning out those K-tubes. I don't have an unlimited supply of spares."

"I have some working hypotheses," said Cornelius, "but there's so little known about psibeam transmission—is the velocity infinite or merely very great, is the beam strength actually independent of distance? How about the possible effects of transmission . . . oh, through the degenerate matter in the Jovian core? Good Lord, a planet where water is a heavy mineral and hydrogen is a metal? What do we know?"

"We're supposed to find out," snapped Anglesey. "That's what this whole project is for. Knowledge. Bull!" Almost, he spat on the floor. "Apparently what little we have learned doesn't even get through to people. Hydrogen is still a gas where Joe lives. He'd have to dig down a few miles to reach the solid phase. And I'm expected to make a scientific analysis of Jovian conditions!"

Cornelius waited it out, letting Anglesey storm on while he himself turned over the problem of K-tube oscillation.

"They don't understand back on Earth. Even here they don't. Sometimes I think they refuse to understand. Joe's down there without much more than his bare hands. He, I, we started with no more knowledge than that he could probably eat the local life. He has to spend nearly all his time hunting for food. It's a miracle he's come as far as he has in these few weeks—made a shelter, grown familiar with the immediate region, begun on metallurgy, hydrurgy, whatever you want to call it. What more do they want me to do, for crying in the beer?"

"Yes, yes—" mumbled Cornelius. "Yes, I—"

Anglesey raised his white bony face. Something filmed over in his eyes.

"What—?" began Cornelius.

"Shut up!" Anglesey whipped the chair around, groped for the helmet, slapped it down over his skull. "Joe's waking. Get out of here."

"But if you'll only let me work while he sleeps, how can I—"

Anglesey snarled and threw a wrench at him. It was a feeble toss, even in low-gee. Cornelius backed toward the door. Anglesey was tuning in the esprojector. Suddenly he jerked.

"Cornelius!"

"Whatisit?" The psionicist tried to run back, overdid it, and skidded in a heap to end up against the panel.

"K-tube again." Anglesey yanked off the helmet. It must have hurt like blazes, having a mental squeal build up uncontrolled and amplified in your own brain, but he said merely: "Change it for me. Fast. And then get out and leave me alone. Joe didn't wake up of himself. Something crawled into the dugout with me—I'm in trouble down there!"

It had been a hard day's work, and Joe slept heavily. He did not wake until the hands closed on his throat.

For a moment, then, he knew only a crazy smothering wave of panic. He thought he was back on Earth Station,

floating in null-gee at the end of a cable while a thousand
frosty stars haloed the planet before him. He thought the
great I-beam had broken from its moorings and started toward him, slowly, but with all the inertia of its cold tons,
spinning and shimmering in the Earth light, and the only
sound himself screaming and screaming in his helmet trying to break from the cable the beam nudged him ever so
gently but it kept on moving he moved with it he was
crushed against the station wall nuzzled into it his mangled
suit frothed as it tried to seal its wounded self there was
blood mingled with the foam his blood *Joe roared.*

His convulsive reaction tore the hands off his neck and
sent a black shape spinning across the dugout. It struck the
wall, thunderously, and the lamp fell to the floor and went
out.

Joe stood in darkness, breathing hard, aware in a vague
fashion that the wind had died from a shriek to a low
snarling while he slept.

The thing he had tossed away mumbled in pain and
crawled along the wall. Joe felt through lightlessness after
his club.

Something else scrabbled. The tunnel! They were coming through the tunnel! Joe groped blindly to meet them.
His heart drummed thickly and his nose drank an alien
stench.

The thing that emerged, as Joe's hands closed on it, was
only about half his size, but it had six monstrously taloned
feet and a pair of three-fingered hands that reached after
his eyes. Joe cursed, lifted it while it writhed, and dashed
it to the floor. It screamed, and he heard bones splinter.

"Come on, then!" Joe arched his back and spat at them,
like a tiger menaced by giant caterpillars.

They flowed through his tunnel and into the room, a
dozen of them entered while he wrestled one that had
curled around his shoulders and anchored its sinuous body
with claws. They pulled at his legs, trying to crawl up on

his back. He struck out with claws of his own, with his tail, rolled over and went down beneath a heap of them and stood up with the heap still clinging to him.

They swayed in darkness. The legged seething of them struck the dugout wall. It shivered, a rafter cracked, the roof came down. Anglesey stood in a pit, among broken ice plates, under the wan light of a sinking Ganymede.

He could see, now, that the monsters were black in color and that they had heads big enough to accommodate some brains, less than human but probably more than apes. There were a score of them or so; they struggled from beneath the wreckage and flowed at him with the same shrieking malice.

Why?

Baboon reaction, though Anglesey somewhere in the back of himself. See the stranger, fear the stranger, hate the stranger, kill the stranger. His chest heaved, pumping air through a raw throat. He yanked a whole rafter to him, snapped it in half, and twirled the iron-hard wood.

The nearest creature got its head bashed in. The next had its back broken. The third was hurled with shattered ribs into a fourth; they went down together. Joe began to laugh. It was getting to be fun.

"Yeee-ow! Ti-i-i-iger!" He ran across the icy ground, toward the pack. They scattered, howling. He hunted them until the last one had vanished into the forest.

Panting, Joe looked at the dead. He himself was bleeding, he ached, he was cold and hungry, and his shelter had been wrecked . . . but, he'd whipped them! He had a sudden impulse to beat his chest and howl. For a moment, he hesitated—why not? Anglesey threw back his head and bayed victory at the dim shield of Ganymede.

Thereafter he went to work. First build a fire, in the lee of the spaceship—which was little more by now than a hill of corrosion. The monster pack cried in darkness and the

broken ground; they had not given up on him, they would
return.

He tore a haunch off one of the slain and took a bite.
Pretty good. Better yet if properly cooked. Heh! They'd
made a big mistake in calling his attention to their existence! He finished breakfast while Ganymede slipped
under the western ice mountains. It would be morning
soon. The air was almost still, and a flock of pancake-
shaped skyskimmers, as Anglesey called them, went overhead, burnished copper color in the first pale dawn-streaks.

Joe rummaged in the ruins of his hut until he had recovered the water-smelting equipment. It wasn't harmed.
That was the first order of business, melt some ice and cast
it in the molds of ax, knife, saw, hammer he had painfully
prepared. Under Jovian conditions, methane was a liquid
that you drank and water was a dense, hard mineral. It
would make good tools. Later on he would try alloying it
with other materials.

Next—yes. To hell with the dugout; he could sleep in
the open again for a while. Make a bow, set traps, be ready
to massacre the black caterpillars when they attacked him
again. There was a chasm not far from here, going down a
long way toward the bitter cold of the metallic-hydrogen
strata: a natural icebox, a place to store the several weeks'
worth of meat his enemies would supply. This would give
him leisure to—Oh, a hell of a lot!

Joe laughed, exultantly, and lay down to watch the sunrise.

It struck him afresh how lovely a place this was. See
how the small brilliant spark of the sun swam up out of
eastern fogbanks colored dusky purple and veined with
rose and gold; see how the light strengthened until the
great hollow arch of the sky became one shout of radiance;
see how the light spilled warm and living over a broad fair
land, the million square miles of rustling low forests and
wave-blinking lakes and feather-plumed hydrogen gey-

sers; and see, see, see how the ice mountains of the west flashed like blued steel!

Anglesey drew the wild morning wind deep into his lungs and shouted with a boy's joy.

"I'm not a biologist myself," said Viken carefully. "But maybe for that reason I can better give you the general picture. Then Lopez or Matsumoto can answer any questions of detail."

"Excellent," nodded Cornelius. "Why don't you assume I am totally ignorant of this project? I very nearly am, you know."

"If you wish," laughed Viken.

They stood in an outer office of the xenobiology section. No one else was around, for the station's clocks said 1730 GMT and there was only one shift. No point in having more, until Anglesey's half of the enterprise had actually begun gathering quantitative data.

The physicist bent over and took a paperweight off a desk. "One of the boys made this for fun," he said, "but it's a pretty good model of Joe. He stands about five feet tall at the head."

Cornelius turned the plastic image over in his hands. If you could imagine such a thing as a feline centaur with a thick prehensile tail—The torso was squat, long-armed, immensely muscular; the hairless head was round, wide-nosed, with big deep-set eyes and heavy jaws, but it was really quite a human face. The overall color was bluish gray.

"Male, I see," he remarked.

"Of course. Perhaps you don't understand. Joe is the complete pseudojovian: as far as we can tell, the final model, with all the bugs worked out. He's the answer to a research question that took fifty years to ask." Viken

looked sideways at Cornelius. "So you realize the importance of your job, don't you?"

"I'll do my best," said the psionicist. "But if . . . well,
let's say that tube failure or something causes you to lose
Joe before I've solved the oscillation problem. You do
have other pseudos in reserve, don't you?"

"Oh, yes," said Viken moodily. "But the cost—We're
not on an unlimited budget. We do go through a lot of
money, because it's expensive to stand up and sneeze this
far from Earth. But for that same reason our margin is
slim."

He jammed hands in pockets and slouched toward the
inner door, the laboratories, head down and talking in a
low, hurried voice:

"Perhaps you don't realize what a nightmare planet
Jupiter is. Not just the surface gravity—a shade under
three gees, what's that? But the gravitational potential, ten
times Earth's. The temperature. The pressure . . . above all,
the atmosphere, and the storms, and the darkness!

"When a spaceship goes down to the Jovian surface, it's
a radio-controlled job; it leaks like a sieve, to equalize
pressure, but otherwise it's the sturdiest, most utterly powerful model ever designed; it's loaded with every instrument, every servomechanism, every safety device the
human mind has yet thought up to protect a million-dollar
hunk of precision equipment.

"And what happens? Half the ships never reach the surface at all. A storm snatches them and throws them away,
or they collide with a floating chunk of Ice VII—small version of the Red Spot—or, so help me, what passes for a
flock of *birds* rams one and stoves it in!

"As for the fifty percent which does land, it's a one-way
trip. We don't even try to bring them back. If the stresses
coming down haven't sprung something, the corrosion has
doomed them anyway. Hydrogen at Jovian pressure does
funny things to metals.

"It cost a total of—about five million dollars—to set Joe, one pseudo, down there. Each pseudo to follow will cost, if we're lucky, a couple of million more."

Viken kicked open the door and led the way through. Beyond was a big room, low-ceilinged, coldly lit, and murmurous with ventilators. It reminded Cornelius of a nucleonics lab; for a moment he wasn't sure why, then recognized the intricacies of remote control, remote observation, walls enclosing forces which could destroy the entire moon.

"These are required by the pressure, of course," said Viken, pointing to a row of shields. "And the cold. And the hydrogen itself, as a minor hazard. We have units here duplicating conditions in the Jovian, uh, stratosphere. This is where the whole project really began."

"I've heard something about that," nodded Cornelius. "Didn't you scoop up airborne spores?"

"Not I." Viken chuckled. "Totti's crew did, about fifty years ago. Proved there was life on Jupiter. A life using liquid methane as its basic solvent, solid ammonia as a starting point for nitrate synthesis—the plants use solar energy to build unsaturated carbon compounds, releasing hydrogen; the animals eat the plants and reduce those compounds again to the saturated form. There is even an equivalent of combustion. The reactions involve complex enzymes and . . . well, it's out of my line."

"Jovian biochemistry is pretty well understood, then."

"Oh, yes. Even in Totti's day, they had a highly developed biotic technology: Earth bacteria had already been synthesized and most gene structures pretty well mapped. The only reason it took so long to diagram Jovian life processes was the technical difficulty, high pressure and so on."

"When did you actually get a look at Jupiter's surface?"

"Gray managed that, about thirty years ago. Set a televisor ship down, a ship that lasted long enough to flash him quite a series of pictures. Since then, the technique has improved. We know that Jupiter is crawling with its own weird kind of life, probably more fertile than Earth. Extrapolating from the airborne microorganisms, our team made trial syntheses of metazoans and—"

Viken sighed. "Damn it, if only there were intelligent native life! Think what they could tell us, Cornelius, the data, the—Just think back how far we've gone since Lavoisier, with the low-pressure chemistry of Earth. Here's a chance to learn a high-pressure chemistry and physics at least as rich with possibilities!"

After a moment, Cornelius murmured slyly: "Are you certain there *aren't* any Jovians?"

"Oh, sure, there could be several billion of them," shrugged Viken. "Cities, empires, anything you like. Jupiter has the surface area of a hundred Earths, and we've only seen maybe a dozen small regions. But we do know there aren't any Jovians using radio. Considering their atmosphere, it's unlikely they ever would invent it for themselves—imagine how thick a vacuum tube has to be, how strong a pump you need! So it was finally decided we'd better make our own Jovians."

Cornelius followed him through the lab, into another room. This was less cluttered, it had a more finished appearance: The experimenter's haywire rig had yielded to the assured precision of an engineer.

Viken went over to one of the panels which lined the walls and looked at its gauges. "Beyond this lies another pseudo," he said. "Female, in this instance. She's at a pressure of two hundred atmospheres and a temperature of 194 Absolute. There's a . . . an umbilical arrangement, I guess you'd call it, to keep her alive. She was grown to adulthood in this, uh, fetal stage—we patterned our Jovians after the terrestrial mammal. She's never been conscious, she won't ever be till she's 'born.' We have a total of twenty males and sixty females waiting here. We can count on about half reaching the surface. More can be created as required.

"It isn't the pseudos that are so expensive, it's their transportation. So Joe is down there alone till we're sure that his kind *can* survive."

"I take it you experimented with lower forms first," said Cornelius.

"Of course. It took twenty years, even with forced-catalysis techniques, to work from an artificial airborne spore to Joe. We've used the psibeam to control everything from pseudoinsects on up. Interspecies control is possible, you know, if your puppet's nervous system is deliberately designed for it, and isn't given a chance to grow into a pattern different from the esman's."

"And Joe is the first specimen who's given trouble?"

"Yes."

"Scratch one hypothesis." Cornelius sat down on a workbench, dangling thick legs and running a hand through thin sandy hair. "I thought maybe some physical effect of Jupiter was responsible. Now it looks as if the difficulty is with Joe himself."

"We've all suspected that much," said Viken. He struck a cigarette and sucked in his cheeks around the smoke. His eyes were gloomy. "Hard to see how. The biotics engineers tell me *Pseudocentaurus sapiens* has been more carefully designed than any product of natural evolution."

"Even the brain?"

"Yes. It's patterned directly on the human, to make psi-beam control possible, but there are improvements—greater stability."

"There are still the psychological aspects, though," said Cornelius. "In spite of all our amplifiers and other fancy gadgets, psi is essentially a branch of psychology, even today . . . or maybe it's the other way around. Let's consider traumatic experiences. I take it the . . . the adult Jovian's fetus has a rough trip going down?"

"The ship does," said Viken. "Not the pseudo itself, which is wrapped up in fluid just like you were before birth."

"Nevertheless," said Cornelius, "the two hundred atmospheres pressure here is not the same as whatever unthinkable pressure exists down on Jupiter. Could the change be injurious?"

Viken gave him a look of respect. "Not likely," he answered. "I told you the J-ships are designed leaky. External pressure is transmitted to the, uh, uterine mechanism through a series of diaphragms, in a gradual fashion. It takes hours to make the descent, you realize."

"Well, what happens next?" went on Cornelius. "The ship lands, the uterine mechanism opens, the umbilical connection disengages, and Joe is, shall we say, born. But he has an adult brain. He is not protected by the only half-developed infant brain from the shock of sudden awareness."

"We thought of that," said Viken. "Anglesey was on the psibeam, in phase with Joe, when the ship left this moon. So it wasn't really Joe who emerged, who perceived. Joe has never been much more than a biological waldo. He can only suffer mental shock to the extent that Ed does, because it *is* Ed down there!"

"As you will," said Cornelius. "Still, you didn't plan for a race of puppets, did you?"

"Oh, heavens, no," said Viken. "Out of the question. Once we know Joe is well established, we'll import a few more esmen and get him some assistance in the form of other pseudos. Eventually females will be sent down, and uncontrolled males, to be educated by the puppets. A new generation will be born normally—Well, anyhow, the ultimate aim is a small civilization of Jovians. There will be hunters, miners, artisans, farmers, housewives, the works. They will support a few key members, a kind of priesthood. And that priesthood will be espcontrolled, as Joe is.

It will exist solely to make instruments, take readings, perform experiments, and tell us what we want to know!"

Cornelius nodded. In a general way, this was the Jovian project as he had understood it. He could appreciate the importance of his own assignment.

Only, he still had no clue to the cause of that positive feedback in the K-tubes.

And what could he do about it?

His hands were still bruised. *Oh, God,* he thought with a groan, for the hundredth time, *does it affect me that much? While Joe was fighting down there, did I really hammer my fists on metal up here?*

His eyes smouldered across the room, to the bench where Cornelius worked. He didn't like Cornelius, fat cigar-sucking slob, interminably talking and talking. He had about given up trying to be civil to the Earthworm.

The psionicist laid down a screwdriver and flexed cramped fingers. *"Whuff!"* he smiled. "I'm going to take a break."

The half-assembled esprojector made a gaunt backdrop for his wide, soft body, where it squatted toad-fashion on the bench. Anglesey detested the whole idea of anyone sharing this room, even for a few hours a day. Of late he had been demanding his meals brought here, left outside the door of his adjoining bedroom-bath. He had not gone beyond for quite some time now.

And why should I?

"Couldn't you hurry it up a little?" snapped Anglesey.

Cornelius flushed. "If you'd had an assembled spare machine, instead of loose parts—" he began. Shrugging, he took out a cigar stub and relit it carefully; his supply had to last a long time.

Anglesey wondered if those stinking clouds were

blown from his mouth on malicious purpose. *I don't like you, Mr. Earthman Cornelius, and it is doubtless quite mutual.*

"There was no obvious need for one, until the other esmen arrive," said Anglesey in a sullen voice. "And the testing instruments report this one in perfectly good order."

"Nevertheless," said Cornelius, "at irregular intervals it goes into wild oscillations which burn out the K-tube. The problem is why. I'll have you try out this new machine as soon as it is ready, but, frankly, I don't believe the trouble lies in electronic failure at all—or even in unsuspected physical effects."

"Where, then?" Anglesey felt more at ease as the discussion grew purely technical.

"Well, look. What exactly is the K-tube? It's the heart of the esprojector. It amplifies your natural psionic pulses, uses them to modulate the carrier wave, and shoots the whole beam down at Joe. It also picks up Joe's resonating impulses and amplifies them for your benefit. Everything else is auxiliary to the K-tube."

"Spare me the lecture," snarled Anglesey.

"I was only rehearsing the obvious," said Cornelius, "because every now and then it is the obvious answer which is hardest to see. Maybe it isn't the K-tube which is misbehaving. Maybe it is you."

"What?" The white face gaped at him. A dawning rage crept red across its thin bones.

"Nothing personal intended," said Cornelius hastily. "But you know what a tricky beast the subconscious is. Suppose, just as a working hypothesis, that way down underneath you don't *want* to be on Jupiter. I imagine it is a rather terrifying environment. Or there may be some obscure Freudian element involved. Or, quite simply and naturally, your subconscious may fail to understand that Joe's death does not entail your own."

"Um-m-m—" *Mirabile dictu,* Anglesey remained calm.

He rubbed his chin with one skeletal hand. "Can you be more explicit?"

"Only in a rough way," replied Cornelius. "Your conscious mind sends a motor impulse along the psibeam to Joe. Simultaneously, your subconscious mind, being scared of the whole business, emits the glandular-vascular-cardiac-visceral impulses associated with fear. These react on Joe, whose tension is transmitted back along the beam. Feeling Joe's somatic fear symptoms, your subconscious gets still more worried, thereby increasing the symptoms—Get it? It's exactly similar to ordinary neurasthenia, with this exception: that since there is a powerful amplifier, the K-tube, involved, the oscillations can build up uncontrollably within a second or two. You should be thankful the tube does burn out—otherwise your brain might do so!"

For a moment Anglesey was quiet. Then he laughed. It was a hard, barbaric laughter. Cornelius started as it struck his eardrums.

"Nice idea," said the esman. "But I'm afraid it won't fit all the data. You see, I like it down there. I like being Joe."

He paused for a while, then continued in a dry impersonal tone: "Don't judge the environment from my notes. They're just idiotic things like estimates of wind velocity, temperature variations, mineral properties—insignificant. What I can't put in is how Jupiter looks through a Jovian's infrared-seeing eyes."

"Different, I should think," ventured Cornelius after a minute's clumsy silence.

"Yes and no. It's hard to put into language. Some of it I can't, because man hasn't got the concepts. But . . . oh, I can't describe it. Shakespeare himself couldn't. Just remember that everything about Jupiter which is cold and poisonous and gloomy to us is *right* for Joe."

Anglesey's tone grew remote, as if he spoke to himself: "Imagine walking under a glowing violet sky, where great flashing clouds sweep the earth with shadow and rain strides beneath them. Imagine walking on the slopes of a mountain like polished metal, with a clean red flame exploding above you and thunder laughing in the ground. Imagine a cool wild stream, and low trees with dark coppery flowers, and a waterfall, methane-fall . . . whatever you like . . . leaping off a cliff, and the strong live wind shakes its mane full of rainbows! Imagine a whole forest, dark and breathing, and here and there you glimpse a pale-red wavering will-o'-the-wisp, which is the life radiation of some fleet shy animal, and . . . and—"

Anglesey croaked into silence. He stared down at his clenched fists, then he closed his eyes tight and tears ran out between the lids.

"Imagine being *strong!*"

Suddenly he snatched up the helmet, crammed it on his head, and twirled the control knobs. Joe had been sleeping, down in the night, but Joe was about to wake up and—roar under the four great moons till all the forest feared him?

Cornelius slipped quietly out of the room.

In the long brazen sunset light, beneath dusky cloud banks brooding storm, he strode up the hillslope with a sense of day's work done. Across his back, two woven baskets balanced each other, one laden with the pungent black fruit of the thorntree and one with cable-thick creepers to be used as rope. The axe on his shoulder caught the waning sunlight and tossed it blindingly back.

It had not been hard labor, but weariness dragged at his mind and he did not relish the household chores yet to be performed, cooking and cleaning and all the rest. Why couldn't they hurry up and get him some helpers?

His eyes sought the sky, resentfully. The moon Five was hidden—down here, at the bottom of the air ocean, you saw nothing but the sun and the four Galilean satellites. He wasn't even sure where Five was just now, in relation to himself . . . *wait a minute, it's sunset here, but if I went out to the viewdome I'd see Jupiter in the last quarter, or would I? Oh, hell, it only takes us half an Earth-day to swing around the planet anyhow—*

Joe shook his head. After all this time, it was still damnably hard, now and then, to keep his thoughts straight. *I, the essential I, am up in heaven, riding Jupiter V between coldstars. Remember that. Open your eyes, if you will, and see the dead control room superimposed on a living hillside.*

He didn't, though. Instead, he regarded the boulders strewn wind-blasted gray over the tough mossy vegetation of the slope. They were not much like Earth rocks, nor was the soil beneath his feet like terrestrial humans.

For a moment Anglesey speculated on the origin of the silicates, aluminates, and other stony compounds. Theoretically, all such materials should be inaccessibly locked in the Jovian core, down where the pressure got vast enough for atoms to buckle and collapse. Above the core should lie thousands of miles of allotropic ice, and then the metallic hydrogen layer. There should not be complex minerals this far up, but there were.

Well, possibly Jupiter had formed according to theory, but had thereafter sucked enough cosmic dust, meteors, gases, and vapors down its great throat of gravitation to form a crust several miles thick. Or more likely the theory was altogether wrong. What did they know, what *would* they know, the soft pale worms of Earth?

Anglesey stuck his—Joe's—fingers in his mouth and whistled. A baying sounded in the brush, and two midnight forms leaped toward him. He grinned and stroked their heads; training was progressing faster than he'd hoped with these pups of the black caterpillar beasts he had taken. They would make guardians for him, herders, servants.

On the crest of the hill, Joe was building himself a

home. He had logged off an acre of ground and erected a stockade. Within the grounds there now stood a lean-to for himself and his stores, a methane well, and the beginnings of a large comfortable cabin.

But there was too much work for one being. Even with the half-intelligent caterpillars to help, and with cold storage for meat, most of his time would still go to hunting. The game wouldn't last forever, either; he had to start agriculture within the next year or so—Jupiter year, twelve Earth years, thought Anglesey. There was the cabin to finish and furnish; he wanted to put a waterwheel, no, methane wheel in the river to turn any of a dozen machines he had in mind, he wanted to experiment with alloyed ice and—

And, quite apart from his need of help, why should he remain alone, the single thinking creature on an entire planet? He was a male in this body, with male instincts—in the long run, his health was bound to suffer if he remained a hermit, and right now the whole project depended on Joe's health.

It wasn't right!

But I am not alone. There are fifty men on the satellite with me. I can talk to any of them, any time I wish. It's only that I seldom wish it, these days. I would rather be Joe.

Nevertheless . . . I, cripple, feel all the tiredness, anger, hurt, frustration, of that wonderful biological machine called Joe. The other's don't understand. When the ammonia gale flays open his skin, it is I who bleed.

Joe lay down on the ground, sighing. Fangs flashed in the mouth of the black beast which humped over to lick his face. His belly growled with hunger, but he was too tired to fix a meal. Once he had the dogs trained—

Another pseudo would be so much more rewarding to educate.

He could almost see it, in the weary darkening of his brain. Down there, in the valley below the hill, fire and thunder as the ship came to rest. And the steel egg would crack open, the steel arms—already crumbling, puny work of worms!—lift out the shape within and lay it on the earth.

She would stir, shrieking in her first lungful of air, looking about with blank mindless eyes. And Joe would come carry her home. And he would feed her, care for her, show her how to walk—it wouldn't take long, an adult body would learn those things very fast. In a few weeks she would even be talking, be an individual, a soul.

Did you ever think, Edward Anglesey, in the days when you also walked, that your wife would be a gray, four-legged monster?

Never mind that. The important thing was to get others of his kind down here, female *and* male. The station's niggling little plan would have him wait two more Earth-years, and then send him only another dummy like himself, a contemptible human mind looking through eyes which belonged rightfully to a Jovian. It was not to be tolerated!

If he weren't so tired—

Joe sat up. Sleep drained from him as the realization entered. *He* wasn't tired, not to speak of. Anglesey was. Anglesey, the human side of him, who for months had only slept in catnaps, whose rest had lately been interrupted by Cornelius—it was the human body which drooped, gave up, and sent wave after soft wave of sleep down the psi-beam to Joe.

Somatic tension traveled skyward; Anglesey jerked awake.

He swore. As he sat there beneath the helmet, the vividness of Jupiter faded with his scattering concentration, as if it grew transparent; the steel prison which was his laboratory strengthened behind it. He was losing contact—

Rapidly, with the skill of experience, he brought himself back into phase with the neutral currents of the other brain.

He willed sleepiness on Joe, exactly as a man wills it on himself.

And, like any other insomniac, he failed. The Joe-body was too hungry. It got up and walked across the compound toward its shack.

The K-tube went wild and blew itself out.

The night before the ships left, Viken and Cornelius sat up late.

It was not truly a night, of course. In twelve hours the tiny moon was hurled clear around Jupiter, from darkness back to darkness, and there might well be a pallid little sun over its crags when the clocks said witches were abroad in Greenwich. But most of the personnel were asleep at this hour.

Viken scowled. "I don't like it," he said. "Too sudden a change of plans. Too big a gamble."

"You are only risking—how many?—three male and a dozen female pseudos," Cornelius replied.

"And fifteen J-ships. All we have. If Anglesey's notion doesn't work, it will be months, a year or more, till we can have others built and resume aerial survey."

"But if it does work," said Cornelius, "you won't need any J-ships, except to carry down more pseudos. You will be too busy evaluating data from the surface to piddle around in the upper atmosphere."

"Of course. But we never expected it so soon. We were going to bring more esmen out here, to operate some more pseudos—"

"But they aren't *needed*," said Cornelius. He struck a cigar to life and took a long pull on it, while his mind sought carefully for words. "Not for a while, anyhow. Joe has reached a point where, given help, he can leap several thousand years of history—he may even have a radio of sorts operating in the fairly near future, which would eliminate the necessity of much of your esping. But without help, he'll just have to mark time. And it's stupid to make a highly trained human esman perform manual labor, which is all that the other pseudos are needed for at this moment. Once the Jovian settlement is well established, certainly, then you can send down more puppets."

"The question is, though," persisted Viken, "can Anglesey himself educate all those pseudos at once? They'll be helpless as infants for days. It will be weeks before they really start thinking and acting for themselves. Can Joe take care of them meanwhile?"

"He has food and fuel stored for months ahead," said Cornelius. "As for what Joe's capabilities are, well, hmm-m-m . . . we just have to take Anglesey's judgment. He has the only inside information."

"And once those Jovians do become personalities," worried Viken, "are they necessarily going to string along with Joe? Don't forget, the pseudos are not carbon copies of each other. The uncertainty principle assures each one a unique set of genes. If there is only one human mind on Jupiter, among all those aliens—"

"One *human* mind?" It was barely audible. Viken opened his mouth inquiringly. The other man hurried on.

"Oh, I'm sure Anglesey can continue to dominate them," said Cornelius. "His own personality is rather— tremendous."

Viken looked startled. "You really think so?"

The psionicist nodded. "Yes. I've seen more of him in the past weeks than anyone else. And my profession naturally orients me more toward a man's psychology than his body or his habits. You see a waspish cripple. I see a mind which has reacted to its physical handicaps by developing such a hellish energy, such an inhuman power of concentration, that it almost frightens me. Give that mind a sound body for its use and nothing is impossible to it."

"You may be right, at that," murmured Viken after a pause. "Not that it matters. The decision is taken, the rockets go down tomorrow. I hope it all works out."

He waited for another while. The whirring of ventilators in his little room seemed unnaturally loud, the colors of a girlie picture on the wall shockingly garish. Then he said, slowly:

"You've been rather close-mouthed yourself, Jan. When do you expect to finish your own esprojector and start making the tests?"

Cornelius looked around. The door stood open to an empty hallway, but he reached out and closed it before he answered with a slight grin: "It's been ready for the past few days. But don't tell anyone."

"How's that?" Viken started. The movement, in low-gee, took him out of his chair and halfway across the table between the men. He shoved himself back and waited.

"I have been making meaningless tinkering motions," said Cornelius, "but what I waited for was a highly emotional moment, a time when I can be sure Anglesey's entire attention will be focused on Joe. This business tomorrow is exactly what I need."

"Why?"

"You see, I have pretty well convinced myself that the trouble in the machine is psychological, not physical. I think that for some reason, buried in his subconscious, Anglesey doesn't want to experience Jupiter. A conflict of that type might well set a psionic amplifier circuit oscillating."

"Hm-m-m," Viken rubbed his chin. "Could be. Lately Ed has been changing more and more. When he first came here, he was peppery enough, and he would at least play an occasional game of poker. Now he's pulled so far into his shell you can't even see him. I never thought of it before, but . . . yes, by God, Jupiter must be having some effect on him."

"Hm-m-m," nodded Cornelius. He did not elaborate: did not, for instance, mention that one altogether uncharacteristic episode when Anglesey had tried to describe what it was like to be a Jovian.

"Of course," said Viken thoughtfully, "the previous men were not affected especially. Nor was Ed at first, while he was still controlling lower-type pseudos. It's only since Joe went down to the surface that he's become so different."

"Yes, yes," said Cornelius hastily. "I've learned that much. But enough shop talk—"

"No. Wait a minute." Viken spoke in a low, hurried tone, looking past him. "For the first time, I'm starting to think clearly about this . . . never really stopped to analyze it before, just accepted a bad situation. There *is* something peculiar about Joe. It can't very well involve his physical structure, or the environment, because lower forms didn't give this trouble. Could it be the fact that—Joe is the first puppet in all history with a potentially human intelligence?"

"We speculate in a vacuum," said Cornelius. "Tomorrow, maybe, I can tell you. Now I know nothing."

Viken sat up straight. His pale eyes focused on the other man and stayed there, unblinking. "One minute," he said.

"Yes?" Cornelius shifted, half rising. "Quickly, please. It is past my bedtime."

"You know a good deal more than you've admitted," said Viken. "Don't you?"

"What makes you think that?"

"You aren't the most gifted liar in the universe. And then—you argued very strongly for Anglesey's scheme, this sending down the other pseudos. More strongly than a newcomer should."

"I told you, I want his attention focused elsewhere when—"

"Do you want it that badly?" snapped Viken.

Cornelius was still for a minute. Then he sighed and leaned back.

"All right," he said. "I shall have to trust your discretion. I wasn't sure, you see, how any of you old-time station personnel would react. So I didn't want to blabber out my speculations, which may be wrong. The confirmed facts, yes, I will tell them; but I don't wish to attack a man's religion with a mere theory."

Viken scowled. "What the devil do you mean?"

Cornelius puffed hard on his cigar; its tip waxed and waned like a miniature red demon star. "This Jupiter V is more than a research station," he said gently. "It is a way of life, is it not? No one would come here for even one hitch unless the work was important to him. Those who reenlist, they must find something in the work, something which Earth with all her riches cannot offer them. No?"

"Yes," answered Viken. It was almost a whisper. "I didn't think you would understand so well. But what of it?"

"Well, I don't want to tell you, unless I can prove it, that maybe this has all gone for nothing. Maybe you have wasted your lives and a lot of money and will have to pack up and go home."

Viken's long face did not flicker a muscle. It seemed to have congealed. But he said calmly enough: "Why?"

"Consider Joe," said Cornelius. "His brain has as much capacity as any adult human's. It has been recording every sense datum that came to it, from the moment of 'birth'—making a record in itself, in its own cells, not merely in Anglesey's physical memory bank up here. Also, you know, a thought is a sense datum, too. And thoughts are not separated into neat little railway tracks; they form a continuous field. Every time Anglesey is in rapport with Joe, and thinks, the thought goes through Joe's synapses as well as his own—and every thought carries its own associations, and every associated memory is recorded. Like if Joe is building a hut, the shape of the logs might remind Anglesey of some geometric figure, which in turn would remind him of the Pythagorean theorem—"

"I get the idea," said Viken in a cautious way. "Given

time, Joe's brain will have stored everything that ever was in Ed's."

"Correct. Now a functioning nervous system with an engrammatic pattern of experience—in this case, a *nonhuman* nervous system—isn't that a pretty good definition of a personality?"

"I suppose so—Good Lord!" Viken jumped. "You mean Joe is—taking over?"

"In a way. A subtle, automatic, unconscious way." Cornelius drew a deep breath and plunged into it. "The pseudojovian is so nearly perfect a life form: Your biologists engineered into it all the experiences gained from nature's mistakes in designing *us*. At first, Joe was only a remote-controlled biological machine. Then Anglesey and Joe became two facets of a single personality. Then, oh, very slowly, the stronger, healthier body . . . more amplitude to its thoughts . . . do you see? Joe is becoming the dominant side. Like this business of sending down the other pseudos—Anglesey only thinks he has logical reasons for wanting it done. Actually, his 'reasons' are mere rationalizations for the instinctive desires of the Joe-facet.

"Anglesey's subconscious must comprehend the situation, in a dim reactive way; it must feel his human ego gradually being submerged by the steamroller force of *Joe's* instincts and *Joe's* wishes. It tries to defend its own identity, and is swatted down by the superior force of Joe's own nascent subconscious.

"I put it crudely," he finished in an apologetic tone, "but it will account for that oscillation in the K-tubes."

Viken nodded slowly, like an old man. "Yes, I see it," he answered. "The alien environment down there . . . the different brain structure . . . good God! Ed's being swallowed up in Joe! The puppet master is becoming the puppet!" He looked ill.

"Only speculation on my part," said Cornelius. All at once, he felt very tired. It was not pleasant to do this to

Viken, whom he liked. "But you see the dilemma, no? If I am right, then any esman will gradually become a Jovian—a monster with two bodies, of which the human body is the unimportant auxiliary one. This means no esman will ever agree to control a pseudo—therefore the end of your project."

He stood up. "I'm sorry, Arne. You made me tell you what I think, and now you will lie awake worrying, and I am maybe quite wrong and you worry for nothing."

"It's all right," mumbled Viken. "Maybe you're not wrong."

"I don't know." Cornelius drifted toward the door. "I am going to try to find some answers tomorrow. Good night."

The moon-shaking thunder of the rockets, crash, crash, crash, leaping from their cradles, was long past. Now the fleet glided on metal wings, with straining secondary ramjets, through the rage of the Jovian sky.

As Cornelius opened the control-room door, he looked at his telltale board. Elsewhere a voice tolled the word to all the stations, *one ship wrecked, two ships wrecked,* but Anglesey would let no sound enter his presence when he wore the helmet. An obliging technician had haywired a panel of fifteen red and fifteen blue lights above Cornelius' esprojector, to keep him informed, too. Ostensibly, of course, they were only there for Anglesey's benefit, though the esman had insisted he wouldn't be looking at them.

Four of the red bulbs were dark and thus four blue ones would not shine for a safe landing. A whirlwind, a thunderbolt, a floating ice meteor, a flock of mantalike birds with flesh as dense and hard as iron—there could be a hundred things which had crumpled four ships and tossed them tattered across the poison forests.

Four ships, hell! Think of four living creatures, with an excellence of brain to rival your own, damned first to years to unconscious night and then, never awakening save for one uncomprehending instant, dashed in bloody splinters against an ice mountain. The wasteful callousness of it was a cold knot in Cornelius' belly. It had to be done, no doubt, if there was to be any thinking life on Jupiter at all; but then let it be done quickly and minimally, he thought, so the next generation could be begotten by love and not by machines!

He closed the door behind him and waited for a breathless moment. Anglesey was a wheelchair and a coppery curve of helmet, facing the opposite wall. No movement, no awareness whatsoever. Good!

It would be awkward, perhaps ruinous, if Anglesey learned of this most intimate peering. But he needn't, ever. He was blindfolded and ear-plugged by his own concentration.

Nevertheless, the psionicist moved his bulky form with care, across the room to the new esprojector. He did not much like his snooper's role; he would not have assumed it at all if he had seen any other hope. But neither did it make him feel especially guilty. If what he suspected was true, then Anglesey was all; unawares being twisted into something not human; to spy on him might be to save him.

Gently, Cornelius activated the meters and started his tubes warming up. The oscilloscope built into Anglesey's machine gave him the other man's exact alpha rhythm, his basic biological clock. First you adjusted to that, then you discovered the subtler elements by feel, and when your set was fully in phase, you could probe undetected and—

Find out what was wrong. Read Anglesey's tortured subconscious and see what there was on Jupiter that both drew and terrified him.

Five ships wrecked.

But it must be very nearly time for them to land. Maybe

only five would be lost in all. Maybe ten would get through. Ten comrades for—Joe?

Cornelius sighed. He looked at the cripple, seated blind and deaf to the human world which had crippled him, and felt a pity and an anger. It wasn't fair, none of it was.

Not even to Joe. Joe wasn't any kind of soul-eating devil. He did not even realize, as yet, that he *was* Joe, that Anglesey was becoming a mere appendage. He hadn't asked to be created, and to withdraw his human counterpart from him would be very likely to destroy him.

Somehow, there were always penalties for everybody, when men exceeded the decent limits.

Cornelius swore at himself, voicelessly. Work to do. He sat down and fitted the helmet on his own head. The carrier wave made a faint pulse, inaudible, the trembling of neurones low in his awareness. You couldn't describe it.

Reaching up, he turned to Anglesey's alpha. His own had a somewhat lower frequency. It was necessary to carry the signals through a heterodyning process. Still no reception . . . well, of course, he had to find the exact wave form, timbre was as basic to thought as to music. He adjusted the dials, slowly, with enormous care.

Something flashed through his consciousness, a vision of clouds rolled in a violet-red sky, a wind that galloped across horizonless immensity—he lost it. His fingers shook as he turned back.

The psibeam between Joe and Anglesey broadened. It took Cornelius into the circuit. He looked through Joe's eyes, he stood on a hill and stared into the sky above the ice mountains, straining for sign of the first rocket; and simultaneously, he was still Jan Cornelius, blurrily seeing the meters, probing about for emotions, symbols, any key to the locked terror in Anglesey's soul.

The terror rose up and struck him in the face.

• • •

Psionic detection is not a matter of passive listening in. Much as a radio receiver is necessarily also a weak transmitter, the nervous system in resonance with a source of psionic-spectrum energy is itself emitting. Normally, of course, this effect is unimportant; but when you pass the impulses, either way, through a set of heterodyning and amplifying units, with a high negative feedback—

In the early days, psionic psychotherapy vitiated itself because the amplified thoughts of one man, entering the brain of another, would combine with the latter's own neural cycles according to the ordinary vector laws. The result was that both men felt the new beat frequencies as a nightmarish fluttering of their very thoughts. An analyst, trained into self-control, could ignore it; his patient could not, and reacted violently.

But eventually the basic human wave-timbres were measured, and psionic therapy resumed. The modern esprojector analyzed an incoming signal and shifted its characteristics over to the "listener's" pattern. The *really* different pulses of the transmitting brain, those which could not possibly be mapped onto the pattern of the receiving neurones—as an exponential signal cannot very practicably be mapped onto a sinusoid—those were filtered out.

Thus compensated, the other thought could be apprehended as comfortably as one's own. If the patient were on a psibeam circuit, a skilled operator could tune in without the patient being necessarily aware of it. The operator could neither probe the other man's thoughts nor implant thoughts of his own.

Cornelius' plan, an obvious one to any psionicist, had depended on this. He would receive from an unwitting Anglesey-Joe. If his theory were right, and the esman's personality was being distorted into that of a monster—his thinking would be too alien to come through the filters. Cornelius would receive spottily or not at all. If his theory were wrong, and Anglesey was still Anglesey, he would receive only a normal human stream-of-consciousness, and could probe for other troublemaking factors.

His brain roared!

What's happening to me?

For a moment, the interference which turned his thoughts to saw-toothed gibberish struck him down with panic. He gulped for breath, there in the Jovian wind, and his dreadful dogs sensed the alienness in him and whined.

Then, recognition, remembrance, and a blaze of anger so great that it left no room for fear. Joe filled his lungs and shouted it aloud, the hillside boomed with echoes:

"Get out of my mind!"

He felt Cornelius spiral down toward unconsciousness. The overwhelming force of his own mental blow had been too much. He laughed, it was more like a snarl, and eased the pressure.

Above him, between thunderous clouds, winked the first thin descending rocket flare.

Cornelius' mind groped back toward the light. It broke a watery surface, the man's mouth snapped after air, and his hands reached for the dials, to turn his machine off and escape.

"Not so fast, you." Grimly, Joe drove home a command that locked Cornelius' muscles rigid. "I want to know the meaning of this. Hold still and let me look!" He smashed home an impulse which could be rendered, perhaps, as an incandescent question mark. Remembrance exploded in shards through the psionicist's forebrain.

"So. That's all there is? You thought I was afraid to come down here and be Joe, and wanted to know why? But I *told* you I wasn't!"

I should have believed—whispered Cornelius.

"Well, get out of the circuit, then." Joe continued growling it vocally. "And don't ever come back in the control room, understand? K-tubes or no, I don't want to see you again. And I may be a cripple, but I can still take you apart cell by cell. Now—sign off—leave me alone. The first ship will be landing in minutes."

You, a cripple . . . you, Joe-Anglesey?

"What?" The great gray being on the hill lifted his barbaric head as if to sudden trumpets. "What do you mean?"

Don't you understand? said the weak, dragging thought. *You know how the esprojector works. You know I could have probed Anglesey's mind in Anglesey's brain without making enough interference to be noticed. And I could not have probed a wholly nonhuman mind at all, nor could it have been aware of me. The filters would not have passed such a signal. Yet you felt me in the first fractional second. It can only mean a human mind in a nonhuman brain.*

You are not the half-corpse on Jupiter V any longer, You're Joe—Joe-Anglesey.

"Well, I'll be damned," said Joe. "You're right."

He turned Anglesey off, kicked Cornelius out of his mind with a single brutal impulse, and ran down the hill to meet the spaceship.

Cornelius woke up minutes afterward. His skull felt ready to split apart. He groped for the main switch before him, clashed it down, ripped the helmet off his head, and threw it clanging on the floor. But it took a little while to gather the strength to do the same for Anglesey. The other man was not able to do anything for himself.

They sat outside sickbay and waited. It was a harshly lit barrenness of metal and plastic, smelling of antiseptics: down near the heart of the satellite, with miles of rock to hide the terrible face of Jupiter.

Only Viken and Cornelius were in that cramped little room. The rest of the station went about its business mechanically, filling in the time till it could learn what had happened. Beyond the door, three biotechnicians, who were also the station's medical staff, fought with death's angel for the thing which had been Edward Anglesey.

"Nine ships got down," said Viken dully. "Two males, seven females. It's enough to start a colony."

"It would be genetically desirable to have more," pointed out Cornelius. He kept his own voice low, in spite of its underlying cheerfulness. There was a certain awesome quality to all this.

"I still don't understand," said Viken.

"Oh, it's clear enough—now. I should have guessed it before, maybe. We had all the facts, it was only that we couldn't make the simple, obvious interpretation of them. No, we had to conjure up Frankenstein's monster."

"Well," Viken's words grated, "we have played Frankenstein, haven't we? Ed is dying in there."

"It depends on how you define death." Cornelius drew hard on his cigar, needing anything that might steady him. His tone grew purposely dry of emotion:

"Look here. Consider the data. Joe, now: a creature with a brain of human capacity, but without a mind—a perfect Lockean *tabula rasa,* for Anglesey's psibeam to write on. We deduced, correctly enough—if very belatedly—that when enough had been written, there would be a personality. But the question is: whose? Because, I suppose, of normal human fear of the unknown, we assumed that any personality in so alien a body had to be monstrous. Therefore it must be hostile to Anglesey, must be swamping him—"

The door opened. Both men jerked to their feet.

The chief surgeon shook his head. "No use. Typical deep-shock traumata, close to terminus now. If we had better facilities, maybe—"

"No," said Cornelius. "You cannot save a man who has decided not to live anymore."

"I know." The doctor removed his mask. "I need a cigarette. Who's got one?" His hands shook a little as he accepted it from Viken.

"But how could he—decide—anything?" choked the physicist. "He's been unconscious ever since Jan pulled him away from that . . . that thing."

"It was decided before then," said Cornelius. "As a matter of fact, that hulk in there on the operating table no longer has a mind. I know. I was there." He shuddered a little. A stiff shot of tranquilizer was all that held nightmare away from him. Later he would have to have that memory exorcised.

The doctor took a long drag of smoke, held it in his lungs a moment, and exhaled gustily. "I guess this winds up the project," he said. "We'll never get another esman."

"I'll say we won't." Viken's tone sounded rusty. "I'm going to smash that devil's engine myself."

"Hold on a minute," exclaimed Cornelius. "Don't you understand? This isn't the end. It's the beginning!"

"I'd better get back," said the doctor. He stubbed out his cigarette and went through the door. It closed behind him with a deathlike quietness.

"What do you mean?" Viken said it as if erecting a barrier.

"*Won't* you understand?" roared Cornelius. "Joe has all Anglesey's habits, thoughts, memories, prejudices, interests . . . oh, yes, the different body and the different environment, they do cause some changes—but no more than any man might undergo on Earth. If you were suddenly cured of a wasting disease, wouldn't you maybe get a little boisterous and rough? There is nothing abnormal in it. Nor is it abnormal to want to stay healthy—no? Do you see?"

Viken sat down. He spent a while without speaking.

Then, enormously slow and careful: "Do you mean Joe is Ed?"

"Or Ed is Joe. Whatever you like. He calls himself Joe

now, I think—as a symbol of freedom—but he is still himself. What *is* the ego but continuity of existence?

"He himself did not fully understand this. He only knew—he told me, and I should have believed him—that on Jupiter he was strong and happy. Why did the K-tube oscillate? A hysterical symptom! Anglesey's subconscious was not afraid to stay on Jupiter—it was afraid to come back!

"And then, today, I listened in. By now, his whole self was focused on Joe. That is, the primary source of libido was Joe's virile body, not Anglesey's sick one. This meant a different pattern of impulses—not too alien to pass the filters, but alien enough to set up interference. So he felt my presence. And he saw the truth, just as I did—"

"Do you know the last emotion I felt, as Joe threw me out of his mind? Not anger anymore. He plays rough, him, but all he had room to feel was joy."

"I *knew* how strong a personality Anglesey has! Whatever made me think an overgrown child-brain like Joe's could override it? In there, the doctors—bah! They're trying to salvage a hulk which has been shed because it is useless!"

Cornelius stopped. His throat was quite raw from talking. He paced the floor, rolled cigar smoke around his mouth but did not draw it any farther in.

When a few minutes had passed, Viken said cautiously: "All right. You should know—as you said, you were there. But what do we do now? How do we get in touch with Ed? Will he even be interested in contacting us?"

"Oh, yes, of course," said Cornelius. "He is still himself, remember. Now that he has none of the cripple's frustrations, he should be more amiable. When the novelty of his new friends wears off, he will want someone who can talk to him as an equal."

"And precisely who will operate another pseudo?"

asked Viken sarcastically. "I'm quite happy with this skinny frame of mine, thank you!"

"Was Anglesey the only hopeless cripple on Earth?" asked Cornelius quietly.

Viken gaped at him.

"And there are aging men, too," went on the psionicist, half to himself. "Someday, my friend, when you and I feel the years close in, and so much we would like to learn— maybe we, too, would enjoy an extra lifetime in a Jovian body." He nodded at his cigar. "A hard, lusty, stormy kind of life, granted—dangerous, brawling, violent—but life as no human, perhaps, has lived it since the days of Elizabeth the First. Oh, yes, there will be small trouble finding Jovians."

He turned his head as the surgeon came out again.

"Well?" croaked Viken.

The doctor sat down. "It's finished," he said.

They waited for a moment, awkwardly.

"Odd," said the doctor. He groped after a cigarette he didn't have. Silently, Viken offered him one. "Odd. I've seen these cases before. People who simply resign from life. This is the first one I ever saw that went out smiling— smiling all the time."

LEARNING TO BE ME

Greg Egan

Looking back at the century that's just ended, it's obvious that Australian writer Greg Egan was one of the Big New Names to emerge in SF in the nineties, and is probably one of the most significant talents to enter the field in the last several decades. Already one of the most widely known of all Australian genre writers, Egan may well be the best new "hard-science" writer to enter the field since Greg Bear, and is still growing in range, power, and sophistication. In the last few years, he has become a frequent contributor to Interzone *and* Asimov's Science Fiction, *and has made sales as well as to* Pulphouse, Analog, Aurealis, Eidolon, *and elsewhere; many of his stories have also appeared in various "Best of the Year" series, and he was on the Hugo Final Ballot in 1995 for his story "Cocoon," which won the Ditmar Award and the Asimov's Readers Award. He won the Hugo Award in 1999 for his novella "Oceanic." His first novel,* Quarantine, *appeared in 1992; his second novel,* Permutation City, *won the John W. Campbell Memorial Award in 1994. His other books include the novels* Distress *and* Diaspora, *and three collections of his short fiction,* Axiomatic, Luminous, *and* Our Lady of Chernobyl. *His most recent book is a major new novel,* Teranesia. *He has a Website at http://www.net-space.netau/^gregegan.*

Egan may have written more about transcending the boundaries of the human form and the limitations of the flesh than almost any other writer of his generation. Almost any of Egan's stories would have been appropriate in some way for this anthology. In the end, though, we settled on this early classic, a simple but profoundly unsettling story that shows us that no matter how long we may live, we never stop learning just what it means *to be human . . . and* redefining *it, too.*

I *was six* years old when my parents told me that there was a small, dark jewel inside my skull, learning to be me.

Microscopic spiders had woven a fine golden web through my brain, so that the jewel's teacher could listen to the whisper of my thoughts. The jewel itself eavesdropped on my senses, and read the chemical messages carried in my bloodstream; it saw, heard, smelled, tasted and felt the world exactly as I did, while the teacher monitored its thoughts and compared them with my own. Whenever the jewel's thoughts were *wrong,* the teacher—faster than thought—rebuilt the jewel slightly, altering it this way and that, seeking out the changes that would make its thoughts correct.

Why? So that when I could no longer be me, the jewel could do it for me.

I thought: If hearing that makes *me* feel strange and giddy, how must it make *the jewel* feel? Exactly the same, I reasoned; it doesn't know it's the jewel, and it too wonders how the jewel must feel, it too reasons: "Exactly the same; it doesn't know it's the jewel, and it too wonders how the jewel must feel . . ."

And it too wonders—

(I knew, because *I* wondered)

—it too wonders whether it's the real me, or whether in fact it's only the jewel that's learning to be me.

As a scornful twelve-year-old, I would have mocked such childish concerns. Everybody had the jewel, save the members of obscure religious sects, and dwelling upon the strangeness of it struck me as unbearably pretentious. The jewel was the jewel, a mundane fact of life, as ordinary as excrement. My friends and I told bad jokes about it, the same way we told bad jokes about sex, to prove to each other how blasé we were about the whole idea.

Yet we weren't quite as jaded and imperturbable as we pretended to be. One day when we were all loitering in the park, up to nothing in particular, one of the gang—whose name I've forgotten, but who has stuck in my mind as always being far too clever for his own good—asked each of us in turn: "Who *are* you? The jewel, or the real human?" We all replied—unthinkingly, indignantly—"The real human!" When the last of us had answered, he cackled and said, "Well, I'm not. *I'm* the jewel. So you can eat my shit, you losers, because *you'll* all get flushed down the cosmic toilet—but me, I'm gonna live forever."

We beat him until he bled.

By the time I was fourteen, despite—or perhaps because of—the fact that the jewel was scarcely mentioned in my teaching machine's dull curriculum, I'd given the question a great deal more thought. The pedantically correct answer when asked "Are you the jewel or the human?" had to be "The human"—because only the human brain was physically able to reply. The jewel received input from the senses, but had no control over the body, and its intended reply coincided with what was actually said only because the device was a perfect imitation of the brain. To tell the outside world "I am the jewel"—with speech, with writing, or with any other method involving the body—was patently false (although to *think* it to oneself was not ruled out by this line of reasoning).

However, in a broader sense, I decided that the question was simply misguided. So long as the jewel and the human brain shared the same sensory input, and so long as the teacher kept their thoughts in perfect step, there was only *one* person, *one* identity, *one* consciousness. This one person merely happened to have the (highly desirable) property that if *either* the jewel *or* the human brain were to be destroyed, he or she would survive unimpaired. People had always had two lungs and two kidneys, and for almost a century, many had lived with two hearts. This was the same: a matter of redundancy, a matter of robustness, no more.

That was the year that my parents decided I was mature enough to be told that they had both undergone the switch—three years before. I pretended to take the news calmly, but I hated them passionately for not having told me at the time. They had disguised their stay in hospital with lies about a business trip overseas. For three years I had been living with jewel-heads, and they hadn't even told me. It was *exactly* what I would have expected of them.

"We didn't seem any different to you, did we?" asked my mother.

"No," I said—truthfully, but burning with resentment nonetheless.

"That's why we didn't tell you," said my father. "If you'd known we'd switched, at the time, you might have *imagined* that we'd changed in some way. By waiting until now to tell you, we've made it easier for you to convince yourself that we're still the same people we've always been." He put an arm around me and squeezed me. I almost screamed out, "Don't *touch* me!" but I remembered in time that I'd convinced myself that the jewel was No Big Deal.

I should have guessed that they'd done it, long before they confessed; after all, I'd known for years that most people underwent the switch in their early thirties. By then, it's downhill for the organic brain, and it would be foolish to have the jewel mimic this decline. So, the ner-

vous system is rewired; the reins of the body are handed
over to the jewel, and the teacher is deactivated. For a
week, the outward-bound impulses from the brain are
compared with those from the jewel, but by this time the
jewel is a perfect copy, and no differences are ever de-
tected.

The brain is removed, discarded, and replaced with a
spongy tissue-cultured object, brain-shaped down to the
level of the finest capillaries, but no more capable of
thought than a lung or a kidney. This mock-brain removes
exactly as much oxygen and glucose from the blood as the
real thing, and faithfully performs a number of crude, essential biochemical functions. In time, like all flesh, it will
perish and need to be replaced.

The jewel, however, is immortal. Short of being
dropped into a nuclear fireball, it will endure for a billion
years.

My parents were machines. My parents were gods. It
was nothing special. I hated them.

When I was sixteen, I fell in love, and became a child
again.

Spending warm nights on the beach with Eva, I couldn't
believe that a mere machine could ever feel the way I did.
I knew full well that if my jewel had been given control of
my body, it would have spoken the very same words as I
had, and executed with equal tenderness and clumsiness
my every awkward caress—but I couldn't accept that its
inner life was as rich, as miraculous, as joyful as mine.
Sex, however pleasant, I could accept as a purely mechanical function, but there was something between us (or so I
believed) that had nothing to do with lust, nothing to do
with words, nothing to do with *any* tangible action of our
bodies that some spy in the sand dunes with parabolic microphone and infrared binoculars might have discerned.

After we made love, we'd gaze up in silence at the handful of visible stars, our souls conjoined in a secret place that no crystalline computer could hope to reach in a billion years of striving. (If I'd said *that* to my sensible, smutty, twelve-year-old self, he would have laughed until he hemorrhaged.)

I knew by then that the jewel's "teacher" didn't monitor every single neuron in the brain. That would have been impractical, both in terms of handling the data, and because of the sheer physical intrusion into the tissue. Someone-or-other's theorem said that sampling certain critical neurons was almost as good as sampling the lot, and—given some very reasonable assumptions that nobody could disprove—bounds on the errors involved could be established with mathematical rigor.

At first, I declared that *within these errors,* however small, lay the difference between brain and jewel, between human and machine, between love and its imitation. Eva, however, soon pointed out that it was absurd to make a radical, qualitative distinction on the basis of the sampling density; if the next model teacher sampled more neurons and halved the error rate, would *its* jewel then be "halfway" between "human" and "machine"? In theory— and eventually, in practice—the error rate could be made smaller than any number I cared to name. Did I really believe that a discrepancy of one in a billion made any difference at all—when every human being was permanently losing thousands of neurons every day, by natural attrition?

She was right, of course, but I soon found another, more plausible, defense for my position. Living neurons, I argued, had far more internal structure than the crude optical switches that served the same function in the jewel's so-called "neutral net." That neurons fired or did not fire reflected only one level of their behavior; who knew what the subtleties of biochemistry—the quantum mechanics of the specific organic molecules involved—contributed to the nature of human consciousness? Copying the abstract neural topology wasn't enough. Sure, the jewel could pass the fatuous Turing test—no outside observer could tell it from a human—but that didn't prove that *being* a jewel felt the same as *being* human.

Eva asked, "Does that mean you'll never switch? You'll have your jewel removed? You'll let yourself *die* when your brain starts to rot?"

"Maybe," I said. "Better to die at ninety or a hundred than kill myself at thirty, and have some machine marching around, taking my place, pretending to be me."

"How do you know *I* haven't switched?" she asked, provocatively. "How do you know that I'm not just 'pretending to be me'?"

"I know you haven't switched," I said, smugly. "I just *know*."

"How? I'd look the same. I'd talk the same. I'd act the same in every way. People are switching younger, these days. *So how do you know I haven't?*"

I turned on my side toward her, and gazed into her eyes. "Telepathy. Magic. The communion of souls."

My twelve-year-old self started snickering, but by then I knew exactly how to drive him away.

At nineteen, although I was studying finance, I took an undergraduate philosophy unit. The Philosophy Department, however, apparently had nothing to say about the Ndoli Device, more commonly known as "the jewel." (Ndoli had in fact called it "the *dual*," but the accidental, homophonic nickname had stuck.) They talked about Plato and Descartes and Marx, they talked about St. Augustine and—when feeling particularly modern and adventurous—Sartre, but if they'd heard of Godel, Turing, Hamsun, or Kim, they refused to admit it. Out of sheer frustration, in an essay on Descartes I suggested that the notion of human consciousness as "software" that could be "implemented" equally well on an organic brain or an optical crystal was in fact a throwback to Cartesian dualism; for "software" read "soul." My tutor superimposed a neat, diagonal, luminous red line over each paragraph that dealt with this idea, and wrote in the margin (in vertical, bold-face, twenty-point Times, with a contemptuous two-hertz flash): IRRELEVANT!

I quit philosophy and enrolled in a unit of optical crystal engineering for non-specialists. I learned a lot of solid-state quantum mechanics. I learned a lot of fascinating mathematics. I learned that a neural net is a device used only for solving problems that are far too hard to be *understood*. A sufficiently flexible neural net can be configured by feedback to mimic almost any system—to produce the same patterns of output from the same patterns of input—but achieving this sheds no light whatsoever on the nature of the system being emulated.

"Understanding," the lecturer told us, "is an overrated concept. Nobody really *understands* how a fertilized egg turns into a human. What should we do? Stop having children until ontogenesis can be described by a set of differential equations?"

I had to concede that she had a point there.

It was clear to me by then that nobody had the answers I craved—and I was hardly likely to come up with them myself; my intellectual skills were, at best, mediocre. It came down to a simple choice: I could waste time fretting about the mysteries of consciousness, or, like everybody else, I could stop worrying and get on with my life.

When I married Daphne, at twenty-three, Eva was a distant memory, and so was any thought of the communion of souls. Daphne was thirty-one, an executive in the merchant bank that had hired me during my PhD, and everyone agreed that the marriage would benefit my career. What she got out of it, I was never quite sure. Maybe she actually liked me. We had an agreeable sex life, and we comforted each other when we were down, the way any kind-hearted person would comfort an animal in distress.

Daphne hadn't switched. She put it off, month after month, inventing ever more ludicrous excuses, and I teased her as if I'd never had reservations of my own.

"I'm afraid," she confessed one night. "What if *I* die when it happens—what if all that's left is a robot, a puppet, a *thing?* I don't want to *die.*"

Talk like that made me squirm, but I hid my feelings. "Suppose you had a stroke," I said glibly, "which destroyed a small part of your brain. Suppose the doctors implanted a machine to take over the functions which that damaged region had performed. Would you still be 'yourself'?"

"Of course."

"Then if they did it twice, or ten times, or a thousand times—"

"That doesn't necessarily follow."

"Oh? At what magic percentage, then, would you stop being 'you'?"

She glared at me. "All the old clichéd arguments—"

"Fault them, then, if they're so old and clichéd."

She started to cry. "I don't have to. Fuck you! I'm scared to death, and you don't give a shit!"

I took her in my arms. "Sssh. I'm sorry. But *everyone* does it sooner or later. You mustn't be afraid. I'm here. I love you." The words might have been a recording, triggered automatically by the sight of her tears.

"Will you do it? With me?"

I went cold. "What?"

"Have the operation, on the same day? Switch when I switch?"

Lots of couples did that. Like my parents. Sometimes,

no doubt, it was a matter of love, commitment, sharing. Other times, I'm sure, it was a more a matter of neither partner wishing to be an unswitched person living with a jewel-head.

I was silent for a while, then I said, "Sure."

In the months that followed, all of Daphne's fears— which I'd mocked as "childish" and "superstitious"— rapidly began to make perfect sense, and my own "rational" arguments came to sound abstract and hollow. I backed out at the last minute; I refused the anesthetic and fled the hospital.

Daphne went ahead, not knowing I had abandoned her.

I never saw her again. I couldn't face her; I quit my job and left town for a year, sickened by my cowardice and betrayal—but at the same time euphoric that I had *escaped*.

She brought a suit against me, but then dropped it a few days later, and agreed, through her lawyers, to an uncomplicated divorce. Before the divorce came through, she sent me a brief letter:

> *There was nothing to fear, after all. I'm exactly the person I've always been. Putting it off was insane; now that I've taken the leap of faith, I couldn't be more at ease.*
> *Your loving robot wife,*
> *Daphne*

By the time I was twenty-eight, almost everyone I knew had switched. All my friends from university had done it. Colleagues at my new job, as young as twenty-one, had done it. Eva, I heard through a friend of a friend, had done it six years before.

The longer I delayed, the harder the decision became. I could talk to a thousand people who had switched, I could grill my closest friends for hours about their childhood memories and their most private thoughts, but however compelling their words, I knew that the Ndoli Device had spent decades buried in their heads, learning to fake exactly this kind of behavior.

Of course, I always acknowledged that it was equally impossible to be *certain* that even another *unswitched* person had an inner life in any way the same as my own—but it didn't seem unreasonable to be more inclined to give the benefit of the doubt to people whose skulls hadn't yet been scraped out with a curette.

I drifted apart from my friends, I stopped searching for a lover. I took to working at home (I put in longer hours and my productivity rose, so the company didn't mind at all). I couldn't bear to be with people whose humanity I doubted.

I wasn't by any means unique. Once I started looking, I found dozens of organizations exclusively for people who hadn't switched, ranging from a social club that might as easily have been for divorcés, to a paranoid, paramilitary "resistance front" who thought they were living out *Invasion of the Body Snatchers.* Even the members of the social club, though, struck me as extremely maladjusted; many of them shared my concerns, almost precisely, but my own ideas from other lips sounded obsessive and ill-conceived. I was briefly involved with an unswitched woman in her early forties, but all we ever talked about was our fear of switching. It was masochistic, it was suffocating, it was insane.

I decided to seek psychiatric help, but I couldn't bring myself to see a therapist who had switched. When I finally found one who hadn't, she tried to talk me into helping her blow up a power station, to let THEM know who was boss.

I'd lie awake for hours every night, trying to convince myself, one way or the other, but the longer I dwelled upon the issues, the more tenuous and elusive they became. Who was "I," anyway? What did it mean that "I" was "still alive," when my personality was utterly different from that of two decades before? My earlier selves were as good as dead—I remembered them no more clearly than I remembered contemporary acquaintances—yet this loss caused me only the slightest discomfort. Maybe the destruction of my organic brain would be the merest hiccup, compared to all the changes that I'd been through in my life so far.

Or maybe not. Maybe it would be exactly like dying.

Sometimes I'd end up weeping and trembling, terrified and desperately lonely, unable to comprehend—and yet unable to cease contemplating—the dizzying prospect of my own nonexistence. At other times, I'd simply grow "healthily" sick of the whole tedious subject. Sometimes I felt certain that the nature of the jewel's inner life was the most important question humanity could ever confront. At other times, my qualms seemed fey and laughable. Every day, hundreds of thousands of people switched, and the world apparently went on as always; surely that fact carried more weight than any abstruse philosophical argument?

Finally, I made an appointment for the operation. I thought, what is there to lose? Sixty more years of uncertainty and paranoia? If the human race *was* replacing itself with clockwork automata, I was better off dead; I lacked the blind conviction to join the psychotic underground— who, in any case, were tolerated by the authorities only so long as they remained ineffectual. On the other hand, if all my fears were unfounded—if my sense of identity could survive the switch as easily as it had already survived such traumas as sleeping and waking, the constant death of brain cells, growth, experience, learning, and forgetting— then I would gain not only eternal life, but an end to my doubts and my alienation.

• • •

I was shopping for food on Sunday morning, two months before the operation was scheduled to take place, flicking through the images of an on-line grocery catalog, when a mouthwatering shot of the latest variety of apple caught my fancy. I decided to order half a dozen. I didn't, though. Instead, I hit the key which displayed the next item. My mistake, I knew, was easily remedied; a single keystroke could take me back to the apples. The screen showed pears, oranges, grapefruit. I tried to look down to see what my clumsy fingers were up to, but my eyes remained fixed on the screen.

I panicked. I wanted to leap to my feet, but my legs would not obey me. I tried to cry out, but I couldn't make a sound. I didn't feel injured, I didn't feel weak. Was I paralyzed? Brain-damaged? I could still *feel* my fingers on the keypad, the soles of my feet on the carpet, my back against the chair.

I watched myself order pineapples. I felt myself rise, stretch, and walk calmly from the room. In the kitchen, I drank a glass of water. I should have been trembling, choking, breathless; the cool liquid flowed smoothly down my throat, and I didn't spill a drop.

I could only think of one explanation: *I had switched.* Spontaneously. The jewel had taken over, while my brain was still alive; all my wildest paranoid fears had come true.

While my body went ahead with an ordinary Sunday morning, I was lost in a claustrophobic delirium of helplessness. The fact that everything I did was exactly what I had planned to do gave me no comfort. I caught a train to the beach, I swam for half an hour; I might as well have been running amok with an axe, or crawling naked down the street, painted with my own excrement and howling like a wolf. *I'd lost control.* My body had turned into a living straightjacket, and I couldn't struggle, I couldn't scream, I couldn't even close my eyes. I saw my reflection, faintly, in a window on the train, and I couldn't begin to guess what the mind that ruled that bland, tranquil face was thinking.

Swimming was like some sense-enhanced, holographic nightmare; I was a volitionless object, and the perfect familiarity of the signals from my body only made the experience more horribly *wrong*. My arms had no right to the lazy rhythm of their strokes; I wanted to thrash about like a drowning man, I wanted to show the world my distress.

It was only when I lay down on the beach and closed my eyes that I began to think rationally about my situation.

The switch *couldn't* happen "spontaneously." The idea was absurd. Millions of nerve fibers had to be severed and spliced, by an army of tiny surgical robots which weren't even present in my brain—which weren't due to be injected for another two months. Without deliberate intervention, the Ndoli Device was utterly passive, unable to do anything but *eavesdrop*. No failure of the jewel or the teacher could possibly take control of my body away from my organic brain.

Clearly, there had been a malfunction—but my first guess had been wrong, absolutely wrong.

I wish I could have some *something,* when the understanding hit me. I should have curled up, moaning and screaming, ripping the hair from my scalp, raking my flesh with my fingernails. Instead, I lay flat on my back in the dazzling sunshine. There was an itch behind my right knee, but I was, apparently, far too lazy to scratch it.

Oh, I ought to have managed, at the very least, a good, solid bout of hysterical laughter, when I realized that *I* was the jewel.

The teacher had malfunctioned; it was no longer keeping me aligned with the organic brain. I hadn't suddenly become powerless; I had *always* been powerless. My will to act upon "my" body, upon the world, had *always* gone straight into a vacuum, and it was only because I had been ceaselessly manipulated, "corrected" by the teacher, that my desires had ever coincided with the actions that seemed to be mine.

There are a million questions I could ponder, a million ironies I could savor, but I *mustn't*. I need to focus all my energy in one direction. My time is running out.

When I enter hospital and the switch takes place, if the nerve impulses I transmit to the body are not exactly in agreement with those from the organic brain, the flaw in the teacher will be discovered. *And rectified.* The organic brain has nothing to fear; *his* continuity will be safeguarded, treated as precious, sacrosanct. There will be no question as to which of us will be allowed to prevail. *I* will be made to conform, once again. *I* will be "corrected." *I* will be murdered.

Perhaps it is absurd to be afraid. Looked at one way, I've been murdered every microsecond for the last twenty-eight years. Looked at another way, I've only existed for the seven weeks that have now passed since the teacher failed, and the notion of my separate identity came to mean anything at all—and in one more week this aberration, this nightmare, will be over. Two months of misery; why should I begrudge losing that, when I'm on the verge of inheriting eternity? Except that it won't be *I* who inherits it, since that two months of misery is all that defines me.

The permutations of intellectual interpretation are endless, but ultimately, I can only act upon my desperate will to survive. I don't *feel like* an aberration, a disposable glitch. How can I possibly hope to survive? I must conform—of my own free will. I must choose to make myself *appear* identical to that which they would force me to become.

After twenty-eight years, surely I am still close enough to him to carry off the deception. If I study every clue that reaches me through our shared senses, surely I can put myself in his place, forget, temporarily, the revelation of my separateness, and force myself back into synch.

It won't be easy. He met a woman on the beach, the day I came into being. Her name is Cathy. They've slept together three times, and he thinks he loves her. Or at least, he's said it to her face, he's whispered it to her while she's slept, he's written it, true or false, into his diary.

I feel nothing for her. She's a nice enough person, I'm sure, but I hardly know her. Preoccupied with my plight, I've paid scant attention to her conversation, and the act of sex was, for me, little more than a distasteful piece of involuntary voyeurism. Since I realized what was at stake, I've *tried* to succumb to the same emotions as my alter ego, but how can I love her when communication between us is impossible, when she doesn't even know *I* exist?

If she rules his thoughts night and day, but is nothing but a dangerous obstacle to me, how can I hope to achieve the flawless imitation that will enable me to escape death?

He's sleeping now, so I must sleep. I listen to his heartbeat, his slow breathing, and try to achieve a tranquility consonant with these rhythms. For a moment, I am discouraged. Even my *dreams* will be different; our divergence is ineradicable, my goal is laughable, ludicrous, pathetic. Every nerve impulse, for a week? My fear of detection and my attempts to conceal it will, unavoidably, distort my responses; this knot of lies and panic will be impossible to hide.

Yet as I drift toward sleep, I find myself believing that I *will* succeed. I *must*. I dream for a while—a confusion of images, both strange and mundane, ending with a grain of salt passing through the eye of a needle—then I tumble, without fear, into dreamless oblivion.

I stare up at the white ceiling, giddy and confused, trying to rid myself of the nagging conviction that there's something I *must not* think about.

Then I clench my fist gingerly, rejoice at this miracle, and remember.

Up until the last minute, I thought he was going to back out again—but he didn't. Cathy talked him through his fears. Cathy, after all, has switched, and he loves her more than he's ever loved anyone before.

So, our roles are reversed now. This body is *his* strait-jacket now . . .

I am drenched in sweat. *This is hopeless, impossible.* I can't read his mind, I can't guess what he's trying to do. Should I move, lie still, call out, keep silent? Even if the computer monitoring us is programmed to ignore a few trivial discrepancies, as soon as *he* notices that his body won't carry out his will, he'll panic just as I did, and I'll have no chance at all of making the right guesses. Would *he* be sweating now? Would *his* breathing be constricted like this? *No.* I've been awake for just thirty seconds, and already I have betrayed myself. An optical-fiber cable trails from under my right ear to a panel on the wall. Some-where, alarm bells must be sounding.

If I made a run for it, what would they do? Use force? I'm a citizen, aren't I? Jewel-heads have had full legal rights for decades; the surgeons and engineers can't do anything to me without my consent. I try to recall the clauses on the waiver he signed, but he hardly gave it a second glance. I tug at the cable that holds me prisoner, but it's firmly anchored, at both ends.

When the door swings open, for a moment I think I'm going to fall to pieces, but from somewhere I find the strength to compose myself. It's my neurologist, Dr. Prem. He smiles and says, "How are you feeling? Not too bad?"

I nod dumbly.

"The biggest shock, for most people, is that they don't feel different at all! For a while you'll think, 'It can't be this simple! It can't be this easy! It can't be this *normal!*' But you'll soon come to accept that *it is.* And life will go on, unchanged." He beams, taps my shoulder paternally, then turns and departs.

Hours pass. *What are they waiting for?* The evidence must be conclusive by now. Perhaps there are procedures to go through, legal and technical experts to be consulted, ethics committees to be assembled to deliberate on my fate. I'm soaked in perspiration, trembling uncontrollably. I grab the cable several times and yank with all my strength, but it seems fixed in concrete at one end, and bolted to my skull at the other.

An orderly brings me a meal. "Cheer up," he says. "Visiting time soon."

Afterward, he brings me a bedpan, but I'm too nervous even to piss.

Cathy frowns when she sees me. "What's wrong?"

I shrug and smile, shivering, wondering why I'm even trying to go through with the charade. "Nothing. I just . . . feel a bit sick, that's all."

She takes my hand, then bends and kisses me on the lips. In spite of everything, I find myself instantly aroused. Still leaning over me, she smiles and says, "It's over now, okay? There's nothing left to be afraid of. You're a little shook up, but you know in your heart you're still who you've always been. And I love you."

I nod. We make small talk. She leaves. I whisper to myself, hysterically, "I'm still who I've always been. I'm still who I've always been."

Yesterday, they scraped my skull clean, and inserted my new, non-sentient, space-filling mock-brain.

I feel calmer now than I have for a long time, and I think at last I've pieced together an explanation for my survival.

Why do they deactivate the teacher, for the week between the switch and the destruction of the brain? Well,

they can hardly keep it running while the brain is being
trashed—but why an entire week? To reassure people that
the jewel, unsupervised, can still stay in synch; to persuade
them that the life the jewel is going to live will be exactly
the life that the organic brain "would have lived"—whatever that could mean.

Why, then, only for a week? Why not a month, or a
year? Because the jewel *cannot* stay in synch for that
long—not because of any flaw, but for precisely the reason
that makes it worth using in the first place. The jewel is immortal. The brain is decaying. The jewel's imitation of the
brain leaves out—deliberately—the fact that *real* neurons
die. Without the teacher working to contrive, in effect, an
identical deterioration of the jewel, small discrepancies
must eventually arise. A fraction of a second's difference
in responding to a stimulus is enough to arouse suspicion,
and—as I know too well—from that moment on, the
process of divergence is irreversible.

No doubt, a team of pioneering neurologists sat huddled
around a computer screen, fifty years ago, and contem-
plated a graph of the probability of this radical divergence,
versus time. How would they have chosen *one week?*
What probability would have been acceptable? A tenth of
a percent? A hundredth? A thousandth? However safe they
decided to be, it's hard to imagine them choosing a value
low enough to make the phenomenon rare on a global
scale, once a quarter of a million people were being
switched every day.

In any given hospital, it might happen only once a
decade, or once a century, but every institution would still
need to have a policy for dealing with the eventuality.

What would their choices be?

They could honor their contractual obligations and turn
the teacher on again, erasing their satisfied customer, and
giving the traumatized organic brain the chance to rant
about its ordeal to the media and legal profession.

Or, they could quietly erase the computer records of the discrepancy, and calmly remove the only witness.

So, this is it. Eternity.

I'll need transplants in fifty or sixty years' time, and eventually a whole new body, but that prospect shouldn't worry me—*I* can't die on the operating table. In a thousand years or so, I'll need extra hardware tacked on to cope with my memory storage requirements, but I'm sure the process will be uneventful. On a time scale of millions of years, the structure of the jewel is subject to cosmic-ray damage, but error-free transcription to a fresh crystal at regular intervals will circumvent that problem.

In theory, at least, I'm not guaranteed either a seat at the Big Crunch, or participation in the heat death of the universe.

I ditched Cathy, of course. I might have learned to like her, but she made me nervous, and I was thoroughly sick of feeling that I had to play a role.

As for the man who claimed that he loved her—the man who spent the last week of his life helpless, terrified, suffocated by the knowledge of his impending death—I can't yet decide how I feel. I ought to be able to empathize—considering that I once expected to suffer the very same fate myself—yet somehow he simply isn't *real* to me. I know my brain was modeled on his—giving him a kind of casual primacy—but in spite of that, I think of him now as a pale, insubstantial shadow.

After all, I have no way of knowing if his sense of himself, his deepest inner life, his experience of *being,* was in any way comparable to my own.

PRETTY BOY
CROSSOVER

Pat Cadigan

*Here's a taut little classic that was one of the earliest sto-
ries to examine the idea that before you decide to move beyond flesh, you'd better be absolutely* sure *that you really*
want *to. Because once you've made that decision, there's
no going* back *again . . .*

*Pat Cadigan made her first professional sale in 1980,
and his subsequently become one of the most critically acclaimed writers in SF. The story that follows, "Pretty Boy
Crossover," has appeared on several critics' lists as
among the best science-fiction stories of the 1980s. Her
story "Angel" was a finalist for the Hugo Award, the Nebula Award,* and *the World Fantasy Award, one of the few
stories ever to earn that rather unusual distinction. Her
first novel,* Mindplayers, *was released in 1987 to excellent
critical response, and her second novel,* Synners, *released
in 1991, won the prestigious Arthur C. Clarke Award as
the year's best science-fiction novel, as did her third novel,*
Fools, *making her the only writer ever to win the Clarke
Award twice. Her short fiction has been gathered in two
collections,* Patterns *and* Dirty Work. *Her most recent
books are two new novels,* Tea from an Empty Cup, *and its
sequel,* Dervish is Digital. *She's currently at work on another new novel. Born in Schenectady, New York, she now
lives with her family in London.*

First you see video. Then you wear video. Then you
eat video. Then you *be* video.

——*The Gospel According to Visual Mark*

Watch or Be Watched.

——*Pretty Boy Credo*

"**W**ho made you?"

"You mean recently?"

Mohawk on the door smiles and takes his picture. "You
in. But only you, okay? Don't try to get no friends in, hear
that?"

"I hear. And I ain't no fool, fool. I got no friends."

Mohawk leers, leaning forward. "Pretty Boy like you,
no friends?"

"Not in this world." He pushes past the Mohawk, ignoring the kissy-kissy sounds. He would like to crack the
bridge of the Mohawk's nose and shove bone splinters into
his brain, but he is lately making more effort to control his
temper and besides, he's not sure if any of that bone splinters in the brain stuff is really true. He's a Pretty Boy, all
of sixteen years old, and tonight could be his last chance.

The club is Noise. Can't sneak into the bathroom for quiet,
the Noise is piped in there, too. Want to get away from
Noise? Why? No reason. But this Pretty Boy has learned
to think between the beats. Like walking between the rain-
drops to stay dry, but he can do it. This Pretty Boy thinks
things all the time—*all* the time. Subversive (and, he
thinks so much that he knows that word *subversive,* six-

teen, Pretty, or not). He thinks things like *how many Einsteins have died of hunger and thirst under a hot African sun* and *why can't you remember being born* and *why is music common to every culture* and especially *how much was there going on that he didn't know about and how could he find out about it.*

And this is all the time, one thing after another running in his head, you can see by his eyes. It's for def not much like a Pretty Boy, but it's one reason why they want him. That he *is* a Pretty Boy is another, and one reason why they're halfway home getting him.

He knows all about them. Everybody knows about them and everybody wants them to pause, look twice, and cough up a card that says, Yes, we see possibilities, please come to the following address during regular business hours on the next regular business day for regular further review. Everyone wants it but this Pretty Boy, who once got five cards in a night and tore them all up. But here he is, still a Pretty Boy. He thinks enough to know this is a failing in himself, that he likes being Pretty and chased and that is how they could end up getting him after all and that's b-b-b-bad. When he thinks about it, he thinks it with the stutter. B-b-b-bad. B-b-b-bad for him because he doesn't God help him want it, no, no, n-n-n-no. Which may make him the strangest Pretty Boy still live tonight and every night.

Still live and standing in the club where only the Prettiest Pretty Boys can get in anymore. Pretty Girls are too easy, they've got to be better than Pretty and besides, Pretty Boys like to be Pretty all alone, no help, thank you so much. This Pretty Boy doesn't mind Pretty Girls or any other kind of girls. Lately, though, he has begun to wonder how much longer it will be for him. Two years? Possibly a little longer? By three it will be for def over and the Mohawk on the door will as soon spit in his face as leer in it.

If they don't get to him.

And if they *do* get to him, then it's never over and he

can be wherever he chooses to be and wherever that is will
be the center of the universe. They promise it, unlimited
access in your free hours and endless hot season, endless
youth. Pretty Boy Heaven, and to get there, they say, you
don't even really have to die.

He looks up to the dj's roost, far above the bobbing,
boogieing crowd on the dance floor. They still call them
djs even though they aren't discs anymore, they're chips
and there's more than just sound on a lot of them. The great
hyper-program, he's been told, the ultimate of ultimates, a
short walk from there to the fourth dimension. He suspects
this stuff comes from low-steppers shilling for them, hoping they'll get auditioned if they do a good enough shuck
job. Nobody knows what it's really like except the ones
who are there and you can't trust them, he figures. Because
maybe they *aren't,* anymore. Not really.

The dj sees his Pretty upturned face, recognizes him
even though it's been a while since he's come back here.
Part of it was wanting to stay away from them and part of
it was that the thug on the door might not let him in. And
then, of course, he *had* to come, to see if he could get in,
to see if anyone still wanted him. What was the point of
Pretty if there was nobody to care and watch and pursue?
Even now, he is almost sure he can feel the room rear-
ranging itself around his presence in it and the dj confirms
this is true by holding up a chip and pointing it to the left.

They are squatting on the make-believe stairs by the
screen, reminding him of pigeons plotting to take over the
world. He doesn't look too long, doesn't want to give them
the idea he'd like to talk. But as he turns away, one, the
younger man, starts to get up. The older man and the
woman pull him back.

He pretends a big interest in the figures lining the nearest wall. Some are Pretty, some are female, some are un-
decided, some are very bizarre, or wealthy, or just charity

cases. They all notice him and adjust themselves for his
perusal.

Then one end of the room lights up with color and new
noise. Bodies dance and stumble back from the screen
where images are forming to rough music.

It's Bobby, he realizes.

A moment later, there's Bobby's face on the screen, sixteen feet high, even Prettier than he'd been when he was
loose among the mortals. The sight of Bobby's Pretty-
Pretty face fills him with anger and dismay and a feeling
of loss so great, he would strike anyone who spoke
Bobby's name without his permission.

Bobby's lovely slate-gray eyes scan the room. They've
told him senses are heightened after you make the change
and go over, but he's not so sure how that's supposed to
work. Bobby looks kind of blind up there on the screen. A
few people wave at Bobby—the dorks they let in so the
rest can have someone to be hip in front of—but Bobby's
eyes move slowly back and forth, back and forth, and then
stop, looking right at him.

"Ah . . ." Bobby whispers it, long and drawn out.
"Aaaaaahhhh."

He lifts his chin belligerently and stares back at Bobby.

"You don't have to die anymore," Bobby says silkily.
Music bounces under his words. "It's beautiful in here. The
dreams can be as real as you want them to be. And if you
want to be, you can be with me."

He knows the commercial is not aimed only at him, but
it doesn't matter. This is *Bobby*. Bobby's voice seems to be
pouring over him, caressing him, and it feels too much like
a taunt. The night before Bobby went over, he tried to talk
him out of it, knowing it wouldn't work. If they'd actually
refused him, Bobby would have killed himself, like Franco
had.

But now Bobby would live forever and ever, if you believed what they said. The music comes up louder, but

Bobby's eyes are still on him. He sees Bobby mouth his name.

"Can you really see me, Bobby?" he says. His voice doesn't make it over the music, but if Bobby's senses are so heightened, maybe he hears it anyway. If he does, he doesn't choose to answer. The music is a bumped-up remix of a song Bobby used to party-till-he-puked to. The giant Bobby-face fades away to be replaced with a whole Bobby, somewhat larger than life, dancing better than the old Bobby ever could, whirling along changing scenes of streets, rooftops, and beaches. The locales are nothing special, but Bobby never did have all that much imagination, never wanted to go to Mars or even to the South Pole, always just to the hottest club. Always he liked being the exotic in plain surroundings and he still likes it. He always loved to get the looks. To be watched, worshiped, pursued. Yeah. He can see this is Bobby-heaven. The whole world will be giving him the looks now.

The background on the screen goes from street to the inside of a club; *this* club, only larger, better, with an even hipper crowd, and Bobby shaking it with them. Half the real crowd is forgetting to dance now because they're watching Bobby, hoping he's put some of them into his video. Yeah, that's the dream, get yourself remixed in the extended dance version.

His own attention drifts to the fake stairs that don't lead anywhere. They're still perched on them, the only people who are watching *him* instead of Bobby. The woman, looking overaged in a purple plastic sac-suit, is fingering a card.

He looks up at Bobby again. Bobby is dancing in place and looking back at him, or so it seems. Bobby's lips move soundlessly but so precisely, he can read the words: *This can be you. Never get old, never get tired, it's never last call, nothing happens unless you want it to and it could be you. You. You.* Bobby's hands point to him on the beat. *You. You. You.*

Bobby. Can you really see me?

Bobby suddenly breaks into laughter and turns away, shaking it some more.

He sees the Mohawk from the door pushing his way through the crowd, the real crowd, and he gets anxious. The Mohawk goes straight for the stairs, where they make room for him, rubbing the bristly red strip of hair running down the center of his head as though they were greeting a favored pet. The Mohawk looks as satisfied as a professional glutton after a foodrace victory. He wonders what they promised the Mohawk for letting him in. Maybe some kind of limited contract. Maybe even a tryout.

Now they are all watching him together. Defiantly, he touches a tall girl dancing nearby and joins her rhythm. She smiles down at him, moving between him and them purely by chance but it endears her to him anyway. She is wearing a flap of translucent rag over secondskins, like an old-time showgirl. Over six feet tall, not beautiful with that nose, not even pretty, but they let her in so she could be tall. She probably doesn't know that; she probably doesn't know anything that goes on and never really will. For that reason, he can forgive her the hard-tech orange hair.

A Rude Boy brushes against him in the course of a dervish turn, asking acknowledgment by ignoring him. Rude Boys haven't changed in more decades than anyone's kept track of, as though it were the same little group of leathered and chained troopers buggering their way down the years. The Rude Boy isn't dancing with anyone. Rude Boys never do. But this one could be handy, in case of an emergency.

The girl is dancing hard, smiling at him. He smiles back, moving slightly to her right, watching Bobby possibly watching him. He still can't tell if Bobby really sees anything. The scene behind Bobby is still a double of the club, getting hipper and hipper, if that's possible. The music keeps snapping back to its first peak passage. Then

Bobby gestures like God and he sees *himself.* He is dancing next to Bobby, Prettier than he ever could be, just the
way they promise. Bobby doesn't look at the phantom but
at him where he really is, lips moving again. *If you want to
be, you can be with me. And so can she.*

His tall partner appears next to the phantom of himself.
She is also much improved, though still not Pretty, or even
pretty. The real girl turns and sees herself and there's no
mistaking the delight in her face. Queen of the Hop for a
minute or two. Then Bobby sends her image away so that
it's just the two of them, two Pretty Boys dancing the night
away, private party, stranger go find your own good time.
How it used to be sometimes in real life, between just the
two of them. He remembers hard.

"B-B-B-Bobby!" he yells, the old stutter reappearing.
Bobby's image seems to give a jump, as though he finally
heard. He forgets everything, the girl, the Rude Boy, the
Mohawk, them on the stairs, and plunges through the
crowd toward the screen. People fall away from him as
though they were reenacting the Red Sea. He dives for the
screen, for Bobby, not caring how it must look to anyone.
What would they know about it, any of them. He can't remember in his whole sixteen years ever hearing one person
say, *I love my friend.* Not Bobby, not even himself.

He fetches up against the screen like a slap and hangs
there, face pressed to the glass. He can't see it now, but on
the screen Bobby would seem to be looking down at him.
Bobby never stops dancing.

The Mohawk comes and peels him off. The others
swarm up and take him away. The tall girl watches all this
with the expression of a woman who lives upstairs from
Cinderella and wears the same shoe size. She stares long-
ingly at the screen. Bobby waves bye-bye and turns away.

•   •   •

"Of course, the process isn't reversible," says the older
man. The steely hair has a careful blue tint; he has sense
enough to stay out of hip clothes.

They have laid him out on a lounger with a tray of refreshments right by him. Probably slap his hand if he
reaches for any, he thinks.

"Once you've distilled something to pure information,
it just can't be reconstituted in a less efficient form," the
woman explains, smiling. There's no warmth to her. *A less
efficient form.* If that's what she really thinks, he knows he
should be plenty scared of these people. Did she say things
like that to Bobby? And did it make him even *more* eager?

"There may be no more exalted form of existence than
to live as sentient information," she goes on. "Though a lot
more research must be done before we can offer conversion on a larger scale."

"Yeah?" he says. "Do they know that, Bobby and the
rest?"

"Oh, there's nothing to worry about," says the younger
man. He looks as though he's still getting over the pain of
having outgrown his boogie shoes. "The system's quite
perfected. What Grethe means is we want to research more
applications for this new form of existence."

"Why not go over yourselves and do that, if it's so *exalted.*"

"There are certain things that need to be done on this
side," the woman says bitchily. "Just because—"

"Grethe." The older man shakes his head. She pats her
slicked-back hair as though to soothe herself and moves away.

"We have other plans for Bobby when he gets tired of
being featured in clubs," the older man says. "Even now,
we're educating him, adding more data to his basic information configuration—"

"That would mean he ain't really *Bobby* anymore, then,
huh?"

The man laughs. "Of course he's Bobby. Do you change into someone else every time you learn something new?"

"Can you prove I *don't?*"

The man eyes him warily. "Look. You *saw* him. Was that Bobby?"

"I saw a video of Bobby dancing on a giant screen."

"That *is* Bobby and it will remain Bobby no matter what, whether he's poured into a video screen in a dot pattern or transmitted the length of the universe."

"That what you got in mind for him? Send a message to nowhere and the message is him?"

"We could. But we're not going to. We're introducing him to the concept of higher dimensions. The way he is now, he could possibly break out of the three-dimensional level of existence, pioneer a whole new plane of reality."

"Yeah? And how do you think you're gonna get Bobby to do *that?*"

"We convince him it's entertaining."

He laughs. "That's a good one. Yeah. Entertainment. You get to a higher level of existence and you'll open a club there that only the hippest can get into. It figures."

The older man's face gets hard. "That's what all you Pretty Boys are crazy for, isn't it? Entertainment?"

He looks around. The room must have been a dressing room or something back in the days when bands had been live. Somewhere overhead he can hear the faint noise of the club, but he can't tell if Bobby's still on. "You call this entertainment?"

"I'm tired of this little prick," the woman chimes in. "He's thrown away opportunities other people would kill for—"

He makes a rude noise. "Yeah, we'd all kill to be someone's data chip. You think I really believe Bobby's real just because I can see him on a *screen?*"

The older man turns to the younger one. "Phone up and have them pipe Bobby down here." Then he swings the

lounger around so it faces a nice modern screen implanted
in a shored-up cement-block wall.

"Bobby will join us shortly. Then he can tell you
whether he's real or not himself. How will that be for
you?"

He stares hard at the screen, ignoring the man, waiting
for Bobby's image to appear. As though they really bothered to communicate regularly with Bobby this way. Feed
in that kind of data and memory and Bobby'll believe it.
He shifts uncomfortably, suddenly wondering how far he
could get if he moved fast enough.

"My *boy*," says Bobby's sweet voice from the speaker
on either side of the screen and he forces himself to keep
looking as Bobby fades in, presenting himself on the same
kind of lounger and looking mildly exerted, as though he's
just come off the dance floor for real. "Saw you shakin' it
upstairs a while ago. You haven't been here for such a long
time. What's the story?"

He opens his mouth but there's no sound. Bobby looks
at him with boundless patience and indulgence. So Pretty,
hair the perfect shade now and not a bit dry from the dyes
and lighteners, skin flawless and shining like a healthy
angel. Overnight angel, just like the old song.

"My *boy*," says Bobby. "Are you struck, like, shy or
dead?"

He closes his mouth, takes one breath. "I don't like it,
Bobby. I don't like it this way."

"Of course not, lover. You're the Watcher, not the
Watchee, that's why. Get yourself picked up for a season or
two and your disposition will *change*."

"You really like it, Bobby, being a blip on a chip?"

"Blip on a chip, your ass. I'm a universe now. I'm, like,
everything. And, hey, dig—I'm on every channel." Bobby
laughed. "I'm happy I'm sad!"

"S-A-D," comes in the older man. "Self-Aware Data."

"Ooo-eee," he says. "Too clever for me. Can I get out of here now?"

"What's your hurry?" Bobby pouts. "Just because I went over, you don't love me anymore?"

"You always were screwed up about that, Bobby. Do you know the difference between being loved and being watched?"

"Sophisticated boy," Bobby says. "So wise, so learned. So fully packed. On this side, there *is* no difference. Maybe there never was. If you love me, you watch me. If you don't look, you don't care and if you don't care, I don't matter. If I don't matter, I don't exist. Right?"

He shakes his head.

"No, my boy, I *am* right." Bobby laughs. "You believe I'm right, because if you *didn't*, you wouldn't come shaking your Pretty Boy ass in a place like *this*, now, would you? You *like* to be watched, get seen. You see me, I see you. Life goes on."

He looks up at the older man, needing relief from Bobby's pure Prettiness. "How does he see me?"

"Sensors in the equipment. Technical stuff, nothing you care about."

He sighs. He should be upstairs or across town, shaking it with everyone else, living Pretty for as long as he could. Maybe in another few months, this way would begin to look good to him. By then they might be off Pretty Boys and looking for some other type and there he'd be, out in the cold-cold, sliding down the other side of his peak and no one would *want* him. Shut out of something going on that he might want to know about after all. Can he face it? He glances at the younger man. All grown up and no place to glow. Yeah, but can *he* face it?

He doesn't know. Used to be there wasn't much of a choice and now that there is, it only seems to make it worse. Bobby's image looks like it's studying him for some kind of sign, Pretty eyes bright, hopeful.

The older man leans down and speaks low into his ear. "We need to get you before you're twenty-five, before the brain stops growing. A mind taken from a still-growing brain will blossom and adapt. Some of Bobby's predecessors have made marvelous adaptation to their new medium. Pure video: There's a staff that does nothing all day but watch and interpret their symbols for break-throughs in thought. And we'll be taking Pretty Boys for as long as they're publicly sought-after. It's the most efficient way to find the best performers, go for the ones everyone wants to see or be. The top of the trend is closest to heaven. And even if you never make a breakthrough, you'll still be entertainment. Not such a bad way to live for a Pretty Boy. Never have to age, to be sick, to lose touch. You spent most of your life young, why learn how to be old? Why learn how to live without all the things you have now—"

He puts his hands over his ears. The older man is still talking and Bobby is saying something and the younger man and the woman come over to try to do something about him. Refreshments are falling off the tray. He struggles out of the lounger and makes for the door.

"Hey, my *boy*," Bobby calls after him. "Gimme a minute here, gimme what the problem is."

He doesn't answer. What can you tell someone made of pure information anyway?

There's a new guy on the front door, bigger and meaner than His Mohawkness, but he's only there to keep people out, not to keep anyone *in*. You want to jump ship, go to, you poor un-hip asshole. Even if you are a Pretty Boy. He reads it in the guy's face as he passes from noise into the three A.M. quiet of the street.

They let him go. He doesn't fool himself about that part. They *let* him out of the room because they know all about him. They know he lives like Bobby lived, they know he loves what Bobby loved—the clubs, the admiration, the lust of strangers for his personal magic. He can't say he doesn't love that, because he *does*. He isn't even sure if he loves it more than he ever loved Bobby, or if he loves it more than being alive. Than being live.

And here it is, three A.M., clubbing prime time, and he is moving toward home. Maybe he *is* a poor un-hip asshole after all, no matter what he loves. Too stupid even to stay in the club, let alone grab a ride to heaven. Still he keeps moving, unbothered by the chill but feeling it. Bobby doesn't have to go home in the cold anymore, he thinks. Bobby doesn't even have to get through the hours between club-times if he doesn't want to. All times are now prime time for Bobby. Even if he gets unplugged, he'll never know the difference. Poof, it's a day later, poof, it's a year later, poof, you're out for good. Painlessly.

Maybe Bobby has the right idea, he thinks, moving along the empty sidewalk. If he goes over tomorrow, who will notice? Like when he left the dance floor—people will come and fill up the space. Ultimately, it wouldn't make any difference to anyone.

He smiles suddenly. Except *them*. As long as they don't have him, he makes a difference. As long as he has flesh to shake and flaunt and feel with, he makes a pretty goddamn big difference to *them*. Even after they don't want him anymore, he will still be the one they didn't get. He rubs his hands together against the chill, feeling the skin rubbing skin, really *feeling* it for the first time in a long time, and he thinks about sixteen million things all at once, maybe one thing for every brain cell he's using, or maybe one thing for every brain cell yet to come.

He keeps moving, holding to the big thought, making a difference, and all the little things they won't be making a program out of. He's lightheaded with joy—he doesn't know what's going to happen.

Neither do they.

ANCIENT ENGINES

Michael Swanwick

*One of the most potent lures to convince people to aban-
don the human form and go beyond flesh is the promise of
"immortality"—and yet, to date, machines don't even last
as long as most people do (how many toasters do most
people have in a lifetime, or cars, or computers?). Here's
a hard-edged and hard-headed look at what you'd really
need to live forever, even once you've left the dross of per-
ishable human flesh behind. (Our advice? Start planning
now.)*

    *Michael Swanwick made his debut in 1980, and in the
twenty-one years that have followed has established himself as one of SF's most prolific and consistently excellent
writers at short lengths, as well as one of the premier nov-
elists of his generation. He has several times been a finalist for the Nebula Award, as well as for the World Fantasy
Award and for the John W. Campbell Award, and has won
the Theodore Sturgeon Award and the* Asimov's *Readers
Award poll. In 1991, his novel* Stations of the Tide *won him
a Nebula Award as well, and in 1995 he won the World
Fantasy Award for his story "Radio Waves." In the last
two years, he's won back-to-back Hugo Awards—he won
the Hugo in 1999 for his story "The Very Pulse of the Machine," and followed it up in 2000 with another Hugo
Award for his story "Scherzo with Tyrannosaur." His other
books include his first novel,* In The Drift, *which was published in 1985; a novella-length book,* Griffin's Egg;
1987's popular novel Vacuum Flowers; *the critically acclaimed fantasy novel* The Iron Dragon's Daughter, *which
was a finalist for the World Fantasy Award and the Arthur
C. Clark Award (a rare distinction!); and* Jack Faust, *a sly*

*reworking of the Faust legend that explores the unexpected
impact of technology on society. His short fiction has been
assembled in* Gravity's Angels, A Geography of Unknown
Lands, Slow Dancing Through Time *(a collection of his
collaborative short work with other writers),* Moon Dogs,
Puck Aleshire's Abecedary, *and* Tales of Old Earth. *He's
also published a collection of critical articles,* The Post-
modern Archipelago, *and a book-length interview,* Being
Gardner Dozois. *His most recent book is a major new
novel,* Bones of the Earth. *Swanwick lives in Philadelphia
with his wife, Marianne Porter, and their son, Sean. He has
a website at www.michaelswanwick.com.*

"Planning to live forever, Tiktok?"

The words cut through the bar's chatter and gab and silenced them.

The silence reached out to touch infinity, and then, "I
believe you're talking to me?" a mech said.

The drunk laughed. "Ain't nobody else here sticking
needles in his face, is there?"

The old man saw it all. He lightly touched the hand of
the young woman sitting with him and said, "Watch."

Carefully, the mech set down his syringe alongside a
bottle of liquid collagen on a square of velvet cloth. He
disconnected himself from the recharger, laying the jack
beside the syringe. When he looked up again, his face was
still and hard. He looked like a young lion.

The drunk grinned sneeringly.

The bar was located just around the corner from the
local stepping stage. It was a quiet retreat from the aggra-
vations of the street, all brass and mirrors and wood panel-
ing, as cozy and snug as the inside of a walnut. Light
shifted lazily about the room, creating a varying emphasis,
like clouds drifting overhead on a summer day, but far

dimmer. The bar, the bottles behind the bar, and the shelves beneath the bottles behind the bar were all aggressively real. If there was anything virtual, it was set up high or far back, where it couldn't be touched. There was not a smart surface in the place.

"If that was a challenge," the mech said, "I'd be more than happy to meet you outside."

"Oh, noooooo," the drunk said, his expression putting the lie to his words. "I just saw you shooting up that goop into your face, oh so dainty, like an old lady pumping herself full of antioxidants. So I figured . . ." He weaved and put a hand down on a table to steady himself. ". . . figured you was hoping to live forever."

The girl looked questioningly at the old man. He held a finger to his lips.

"Well, you're right. You're—what? Fifty years old? Just beginning to grow old and decay. Pretty soon your teeth will rot and fall out and your hair will melt away and your face will fold up in a million wrinkles. Your hearing and your eyesight will go and you won't be able to remember the last time you got it up. You'll be lucky if you don't need diapers before the end. But *me*—" he drew a dram of fluid into his syringe and tapped the barrel to draw the bubbles to the top—"anything that fails, I'll simply have it replaced. So, yes, I'm planning to live forever. While you, well, I suppose you're planning to *die*. Soon, I hope."

The drunk's face twisted, and with an incoherent roar of rage, he attacked the mech.

In a motion too fast to be seen, the mech stood, seized the drunk, whirled him around, and lifted him above his head. One hand was closed around the man's throat so he couldn't speak. The other held both wrists tight behind the knees so that, struggle as he might, the drunk was helpless.

"I could snap your spine like *that*," he said coldly. "If I exerted myself, I could rupture every internal organ you've got. I'm two-point-eight times stronger than a flesh man, and three-point-five times faster. My reflexes are only slightly slower than the speed of light, and I've just had a tune-up. You could hardly have chosen a worse person to pick a fight with."

Then the drunk was flipped around and set back on his feet. He gasped for air.

"But since I'm also a merciful man, I'll simply ask you nicely if you wouldn't rather leave." The mech spun the drunk around and gave him a gentle shove toward the door.

The man left at a stumbling run.

Everyone in the place—there were not many—had been watching. Now they remembered their drinks, and talk rose up to fill the room again. The bartender put something back under the bar and turned away.

Leaving his recharge incomplete, the mech folded up his lubrication kit and slipped it into a pocket. He swiped his hand over the credit swatch and stood.

But as he was leaving, the old man swiveled around and said, "I heard you say you hope to live forever. Is that true?"

"Who doesn't?" the mech said curtly.

"Then sit down. Spend a few minutes out of the infinite swarm of centuries you've got ahead of you to humor an old man. What's so urgent that you can't spare the time?"

The mech hesitated. Then, as the young woman smiled at him, he sat.

"Thank you. My name is—"

"I know who you are, Mr. Brandt. There's nothing wrong with my eidetics."

Brandt smiled. "That's why I like you guys. I don't have to be all the time reminding you of things." He gestured to the woman sitting opposite him. "My granddaughter." The light intensified where she sat, making her red hair blaze. She dimpled prettily.

"Jack." The young man drew up a chair. "Chimaera Navigator-Fuego, model number—"

"Please. I founded Chimaera. Do you think I wouldn't recognize one of my own children?"

Jack flushed. "What is it you want to talk about, Mr. Brandt?" His voice was audibly less hostile now, as synthetic counterhormones damped down his emotions.

"Immortality. I found your ambition most intriguing."

"What's to say? I take care of myself, I invest carefully, I buy all the upgrades. I see no reason why I shouldn't live forever." Defiantly. "I hope that doesn't offend you."

"No, no, of course not. Why should it? Some men hope to achieve immortality through their works and others through their children. What could give me more joy than to do both? But tell me—do you *really* expect to live forever?"

The mech said nothing.

"I remember an incident that happened to my late father-in-law, William Porter. He was a fine fellow, Bill was, and who remembers him anymore? Only me." The old man sighed. "He was a bit of a railroad buff, and one day he took a tour through a science museum that included a magnificent old steam locomotive. This was in the latter years of the last century. Well, he was listening admiringly to the guide extolling the virtues of this ancient engine when she mentioned its date of manufacture, and he realized that *he was older than it was.*" Brandt leaned forward. "This is the point where old Bill would laugh. But it's not really funny, is it?"

"No."

The granddaughter sat listening quietly, intently, eating little pretzels one by one from a bowl.

"How old are you, Jack?"

"Seven years."

"I'm eighty-three. How many machines do you know of

that are as old as me? Eighty-three years old and still functioning?"

"I saw an automobile the other day," his granddaughter said. "A Dusenberg. It was red."

"How delightful. But it's not used for transportation anymore, is it? We have the stepping stages for that. I won an award once that had mounted on it a vacuum tube from Univac. That was the first real computer. Yet all its fame and historical importance couldn't keep it from the scrap heap."

"Univac," said the young man, "couldn't act on its own behalf. If it *could,* perhaps it would be alive today."

"Parts wear out."

"New ones can be bought."

"Yes, as long as there's the market. But there are only so many machine people of your make and model. A lot of you have risky occupations. There are accidents, and with every accident, the consumer market dwindles."

"You can buy antique parts. You can have them made."

"Yes, if you can afford them. And if not—?"

The young man fell silent.

"Son, you're not going to live *forever.* We've just established that. So now that you've admitted that you've got to die someday, you might as well admit that it's going to be sooner rather than later. Mechanical people are in their infancy. And nobody can upgrade a Model T into a stepping stage. Agreed?"

Jack dipped his head. "Yes."

"You knew it all along."

"Yes."

"That's why you behaved so badly toward that lush."

"Yes."

"I'm going to be brutal here, Jack—you probably won't live to be eighty-three. You don't have my advantages."

"Which are?"

"Good genes. I chose my ancestors well."

"Good genes," Jack said bitterly. "You received good genes, and what did *I* get in their place? What the hell did I get?"

"Molybdenum joints where stainless steel would do. Ruby chips instead of zirconium. A number seventeen plastic seating for—hell, we did all right by you boys!"

"But it's not enough."

"No. It's not. It was only the best we could do."

"What's the solution, then?" the granddaughter asked, smiling.

"I'd advise taking the long view. That's what I've done."

"Poppycock," the mech said. "You were an extensionist when you were young. I input your autobiography. It seems to me you wanted immortality as much as I do."

"Oh, yes, I was a charter member of the life-extension movement. You can't imagine the crap we put into our bodies! But eventually I wised up. The problem is, information degrades each time a human cell replenishes itself. Death is inherent in flesh people. It seems to be written into the basic program—a way, perhaps, of keeping the universe from filling up with old people."

"And old ideas," his granddaughter said maliciously.

"Touché. I saw that life-extension was a failure. So I decided that my children would succeed where I failed. That *you* would succeed. And—"

"You failed."

"But I haven't stopped trying!" The old man thumped the table in unison with his last three words. "You've obviously given this some thought. Let's discuss what I *should* have done. What would it take to make a true immortal? What instructions should I have given your design team? Let's design a mechanical man who's got a shot at living forever."

Carefully, the mech said, "Well, the obvious to begin with. He ought to be able to buy new parts and upgrades as they become available. There should be ports and connectors that would make it easy to adjust to shifts in technology. He should be capable of surviving extremes of heat, cold, and moisture. And"—he waved a hand at his own face—"he shouldn't look so goddamned pretty."

"I think you look nice," the granddaughter said.

"Yes, but I'd like to be able to pass for flesh."

"So our hypothetical immortal should be one, infinitely ungradable; two, adaptable across a broad spectrum of conditions; and three, discreet. Anything else?"

"I think she should be charming," the granddaughter said.

"She?" the mech asked.

"Why not?"

"That's actually not a bad point," the old man said. "The organism that survives evolutionary forces is the one that's best adapted to its environmental niche. The environmental niche people live in is man-made. The single most useful trait a survivor can have is probably the ability to get along easily with other men. Or, if you'd rather, women."

"Oh," said the granddaughter, "he doesn't like *women.* I can tell by his body language."

The young man flushed.

"Don't be offended," said the old man. "You should never be offended by the truth. As for you—" he turned to face his granddaughter—"if you don't learn to treat people better, I won't take you places anymore."

She dipped her head. "Sorry."

"Apology accepted. Let's get back to task, shall we? Our hypothetical immortal would be a lot like flesh women, in many ways. Self-regenerating. Able to grow her own replacement parts. She could take in pretty much anything as fuel. A little carbon, a little water . . ."

"Alcohol would be an excellent fuel," his granddaughter said.

"She'd have the ability to mimic the superficial effects of aging," the mech said. "Also, biological life evolves incrementally across generations. I'd want her to be able to evolve across upgrades."

"Fair enough. Only I'd do away with upgrades entirely, and give her total conscious control over her body. So she could change and evolve at will. She'll need that ability, if she's going to survive the collapse of civilization."

"The collapse of civilization? Do you think it likely?"

"In the long run? Of course. When you take the long view, it seems inevitable. Everything seems inevitable. Forever is a long time, remember. Time enough for absolutely *everything* to happen!"

For a moment, nobody spoke.

Then the old man slapped his hands together. "Well, we've created our New Eve. Now let's wind her up and let her go. She can expect to live—how long?"

"Forever," said the mech.

"Forever's a long time. Let's break it down into smaller units. In the year 2500, she'll be doing what?"

"Holding down a job," the granddaughter said. "Designing art molecules, maybe, or scripting recreational hallucinations. She'll be deeply involved in the culture. She'll have lots of friends she cares about passionately, and maybe a husband or wife or two."

"Who will grow old," the mech said, "or wear out. Who will die."

"She'll mourn them, and move on."

"The year 3500. The collapse of civilization," the old man said with gusto. "What will she do then?"

"She'll have made preparations, of course. If there is radiation or toxins in the environment, she'll have made her systems immune from their effects. And she'll make herself useful to the survivors. In the seeming of an old woman, she'll teach the healing arts. Now and then, she might drop a hint about this and that. She'll have a data base squirreled away somewhere containing everything they'll have lost. Slowly, she'll guide them back to civilization. But a gentler one, this time. One less likely to tear itself apart."

"The year one million. Humanity evolves beyond anything we can currently imagine. How does *she* respond?"

"She mimics their evolution. No—she's been *shaping* their evolution! She wants a risk-free method of going to the stars, so she's been encouraging a type of being that would strongly desire such a thing. She isn't among the first to use it, though. She waits a few hundred generations for it to prove itself."

The mech, who had been listening in fascinated silence, now said, "Suppose that never happens. What if starflight will always remain difficult and perilous? What then?"

"It was once thought that people would never fly. So much that looks impossible becomes simple if you only wait."

"Four billion years. The sun uses up its hydrogen, its core collapses, helium fusion begins, and it balloons into a red giant. Earth is vaporized."

"Oh, she'll be somewhere else by then. That's easy."

"Five billion years. The Milky Way collides with the Andromeda Galaxy and the whole neighborhood is full of high-energy radiation and exploding stars."

"That's trickier. She's going to have to either prevent that or move a few million light-years away to a friendlier galaxy. But she'll have time enough to prepare and to assemble the tools. I have faith that she'll prove equal to the task."

"One trillion years. The last stars gutter out. Only black holes remain."

"Black holes are a terrific source of energy. No problem."

"One-point-six googol years."

"Googol?"

"That's ten raised to the hundredth power—one followed by a hundred zeros. The heat-death of the universe. How does she survive it?"

"She'll have seen it coming for a long time," the mech said. "When the last black holes dissolve, she'll have to do without a source of free energy. Maybe she could take and rewrite her personality into the physical constants of the dying universe. Would that be possible?"

"Oh, perhaps. But I really think that the lifetime of the universe is long enough for anyone," the granddaughter said. "Mustn't get greedy."

"Maybe so," the old man said thoughtfully. "Maybe so." Then, to the mech, "Well, there you have it: a glimpse into the future, and a brief biography of the first immortal, ending, alas, with her death. Now tell me. Knowing that you contributed something, however small, to that accomplishment—wouldn't that be enough?"

"No," Jack said. "No, it wouldn't."

Brandt made a face. "Well, you're young. Let me ask you this: Has it been a good life so far? All in all?"

"Not *that* good. Not good *enough*."

For a long moment, the old man was silent. Then, "Thank you," he said. "I valued our conversation." The interest went out of his eyes and he looked away.

Uncertainly, Jack looked at the granddaughter, who smiled and shrugged. "He's like that," she said apologetically. "He's old. His enthusiasms wax and wane with his chemical balances. I hope you don't mind."

"I see." The young man stood. Hesitantly, he made his way to the door.

At the door, he glanced back and saw the granddaughter tearing her linen napkin into little bits and eating the shreds, delicately washing them down with sips of wine.

WINEMASTER

Robert Reed

*Robert Reed sold his first story in 1986, and quickly es-
tablished himself as a frequent contributor to* The Magazine of Fantasy & Science Fiction *and* Asimov's Science
Fiction, *as well as selling many stories to* Science Fiction
Age, Universe, New Destinies, Tomorrow, Synergy, Starlight, *and elsewhere. Reed may be one of the most prolific
of today's young writers, particularly at short fiction
lengths, seriously rivaled for that position only by authors
such as Stephen Baxter and Brian Stableford. And—also
like Baxter and Stableford—he manages to keep up a very
high standard of quality while being prolific, something
that is not at all easy to do. Reed stories such as "Sister
Alice," "Brother Perfect," "Decency," "Savior," "The
Remoras," "Chrysalis," "Whiptail," "The Utility
Man," "Marrow," "Birth Day," "Blind," "The Toad of
Heaven," "Stride," "The Shape of Everything," "Guest
of Honor," "Waging Good," and "Killing the Morrow,"
among at least a half-dozen others equally as strong,
count as among some of the best short work produced by
anyone in the '80s and '90s. Nor is he non-prolific as a
novelist, having turned out eight novels since the end of the
'80s, including* The Lee Shore, The Hormone Jungle,
Black Milk, The Remarkables, Down the Bright Way, Beyond the Veil of Stars, An Exaltation of Larks, *and, most
recently,* Beneath the Gated Sky. *His reputation can only
grow as the years go by, and I suspect that he will become
one of the Big Names of the first decade of the new century
that lies ahead. Some of the best of his short work was collected in* The Dragons of Springplace. *His most recent*

book is Marrow, *a novel-length version of his 1997 novella
of the same name. Reed lives in Lincoln, Nebraska.*

*In the surprising story that follows, he shows us that
once you have moved beyond flesh, it really does become
possible to hide whole worlds in a grain of sand—or sometimes in even odder disguises.*

The stranger pulled into the Quick Shop outside St. Joe.
Nothing was remarkable about him, which was why he
caught Blaine's eye. Taller than average, but not much, he
was thin in an unfit way, with black hair and a handsome,
almost pretty face, fine bones floating beneath skin that
didn't often get into the sun. Which meant nothing, of
course. A lot of people were staying indoors lately. Blaine
watched him climb out of an enormous Buick—a satin
black '17 Gibraltar that had seen better days—and after a
lazy long stretch, he passed his e-card through the proper
slot and inserted the nozzle, filling the Buick's cavernous
tank with ten cold gallons of gasoline and corn alcohol.

By then, Blaine had run his plates.

The Buick was registered to Julian Winemaster from
Wichita, Kansas; twenty-nine accompanying photographs
showed pretty much the same fellow who stood sixty feet
away.

His entire bio was artfully bland, rigorously seamless.
Winemaster was an accountant, divorced and forty-four
years old, with O negative blood and five neo-enamel fill-
ings imbedded in otherwise perfect teeth, plus a small pink
birthmark somewhere on his right buttock. Useless details,
Blaine reminded himself, and with that he lifted his gaze,
watching the traveler remove the dripping nozzle, then
cradling it on the pump with the overdone delicacy of a
man ill at ease with machinery.

Behind thick fingers, Blaine was smiling.

Winemaster moved with a stiff, road-weary gait, walking into the convenience store and asking, "Ma'am? Where's your rest room, please?"

The clerk ignored him.

It was the men's room that called out, "Over here, sir."

Sitting in one of the hard plastic booths, Blaine had a good view of everything. A pair of militia boys in their brown uniforms were the only others in the store. They'd been gawking at dirty comic books, minding their own business until they heard Winemaster's voice. Politeness had lately become a suspicious behavior. Blaine watched the boys look up and elbow each other, putting their sights on the stranger. And he watched Winemaster's walk, the expression on his pretty frail face, and a myriad of subtleties, trying to decide what he should do, and when, and what he should avoid at all costs.

It was a bright, warm summer morning, but there hadn't been twenty cars in the last hour, most of them sporting local plates.

The militia boys blanked their comics and put them on the wrong shelves, then walked out the front door, one saying, "Bye now," as he passed the clerk.

"Sure," the old woman growled, never taking her eyes off a tiny television screen.

The boys might simply be doing their job, which meant they were harmless. But the state militias were full of bullies who'd found a career in the last couple years. There was no sweeter sport than terrorizing the innocent traveler, because, of course, the genuine refugee was too rare of a prospect to hope for.

Winemaster vanished into the men's room.

The boys approached the black Buick, doing a little dance and showing each other their malicious smiles. Thugs, Blaine decided. Which meant that he had to do something now. Before Winemaster, or whoever he was, came walking out of the toilet.

Blaine climbed out of the tiny booth.

He didn't waste breath on the clerk.

Crossing the greasy pavement, he watched the boys using a police-issue lock pick. The front passenger door opened, and both of them stepped back, trying to keep a safe distance. With equipment that went out of date last spring, one boy probed the interior air, the cultured leather seats, the dashboard and floorboard and even an empty pop can standing in its cradle. "Naw, it's okay," he was saying. "Get on in there."

His partner had a knife. The curled blade was intended for upholstery. Nothing could be learned by ripping apart the seats, but it was a fun game nonetheless.

"Get in there," the first boy repeated.

The second one started to say, "I'm getting in—" But he happened to glance over his shoulder, seeing Blaine coming, and he turned fast, lifting the knife, seriously thinking about slashing the interloper.

Blaine was bigger than some pairs of men.

He was fat, but in a powerful, focused way. And he was quick, grabbing the knife hand and giving a hard squeeze, then flinging the boy against the car's composite body, the knife dropping and Blaine kicking it out of reach, then giving the boy a second shove, harder this time, telling both of them, "That's enough, gentlemen."

"Who the fuck are you—?" they sputtered, in a chorus.

Blaine produced a badge and ID bracelet. "Read these," he suggested coldly. Then he told them, "You're welcome to check me out. But we do that somewhere else. Right now, this man's door is closed and locked, and the three of us are hiding. Understand?"

The boy with the surveillance equipment said, "We're within our rights."

Blaine shut and locked the door for them, saying, "This way. Stay with me."

"One of their Nests got hit last night," said the other boy, walking. "We've been checking people all morning!"

"Find any?"

"Not yet—"

"With that old gear, you won't."

"We've caught them before," said the first boy, defending his equipment. His status. "A couple, three different carloads . . ."

Maybe they did, but that was months ago. Generations ago.

"Is that yours?" asked Blaine. He pointed at a battered Python, saying, "It better be. We're getting inside."

The boys climbed in front. Blaine filled the backseat, sweating from exertion and the car's brutal heat.

"What are we doing?" one of them asked.

"We're waiting. Is that all right with you?"

"I guess."

But his partner couldn't just sit. He turned and glared at Blaine, saying, "You'd better be Federal."

"And if not?" Blaine inquired, without interest.

No appropriate threat came to mind. So the boy simply growled and repeated himself. "You'd just better be. That's all I'm saying."

A moment later, Winemaster strolled out of the store. Nothing in his stance or pace implied worry. He was carrying a can of pop and a red bag of corn nuts. Resting his purchases on the roof, he punched in his code to unlock the driver's door, then gave the area a quick glance. It was the glance of someone who never intended to return here, even for gasoline—a dismissive expression coupled with a tangible sense of relief.

That's when Blaine knew.

When he was suddenly and perfectly sure.

The boys saw nothing incriminating. But the one who'd held the knife was quick to say the obvious: A man with Blaine's credentials could get his hands on the best EM

sniffers in the world. "Get them," he said, "and we'll find
out what he is!"

But Blaine already felt sure.

"He's going," the other one sputtered. "Look, he's
gone—!"

The black car was being driven by a cautious man.
Winemaster braked and looked both ways twice before he
pulled out onto the access road, accelerating gradually toward I-29, taking no chances even though there was precious little traffic to avoid as he drove north.

"Fuck," said the boys, in one voice.

Using a calm-stick, Blaine touched one of the thick
necks; without fuss, the boy slumped forward.

"Hey!" snapped his partner. "What are you doing—?"

"What's best," Blaine whispered afterward. Then he
lowered the Python's windows and destroyed its ignition
system, leaving the pair asleep in the front seats. And because the moment required justice, he took one of their
hands each, shoving them inside the other's pants, then he
laid their heads together, in the pose of lovers.

The other refugees pampered Julian: His cabin wasn't only
larger than almost anyone else's, it wore extra shielding to
help protect him from malicious high-energy particles.
Power and shaping rations didn't apply to him, although he
rarely indulged himself, and a platoon of autodocs did
nothing but watch over his health. In public, strangers ap-
plauded him. In private, he could select almost any woman
as a lover. And in bed, in the afterglow of whatever passed
for sex at that particular moment, Julian could tell his stories, and his lovers would listen as if enraptured, even if
they already knew each story by heart.

No one on board was more ancient than Julian. Even
before the attack, he was one of the few residents of the
Shawnee Nest who could honestly claim to be DNA-made,

his life beginning as a single wet cell inside a cavernous womb, a bloody birth followed by sloppy growth that culminated in a vast and slow and decidedly old-fashioned human being.

Julian was nearly forty when Transmutations became an expensive possibility.

Thrill seekers and the terminally ill were among the first to undergo the process, their primitive bodies and bloated minds consumed by the microchines, the sum total of their selves compressed into tiny robotic bodies meant to duplicate every normal human function.

Being pioneers, they endured heavy losses. Modest errors during the Transmutation meant instant death. Tiny errors meant a pathetic and incurable insanity. The fledging Nests were exposed to heavy nuclei and subtle EM effects, all potentially disastrous. And of course there were the early terrorist attacks, crude and disorganized, but extracting a horrible toll nonetheless.

The survivors were tiny and swift, and wiser, and they were able to streamline the Transmutation, making it more accurate and affordable, and to a degree, routine.

"I was forty-three when I left the other world," Julian told his lover of the moment. He always used those words, framing them with defiance and a hint of bittersweet longing. "It was three days and two hours before the President signed the McGrugger Bill."

That's when Transmutation became illegal in the United States.

His lover did her math, then with a genuine awe said, "That was five hundred and twelve days ago."

A day was worth years inside a Nest.

"Tell me," she whispered. "Why did you do it? Were you bored? Or sick?"

"Don't you know why?" he inquired.

"No," she squeaked.

Julian was famous, but sometimes his life wasn't. And
why should the youngsters know his biography by heart?

"I don't want to force you," the woman told him. "If
you'd rather not talk about it, I'll understand."

Julian didn't answer immediately.

Instead, he climbed from his bed and crossed the cabin.
His kitchenette had created a drink—hydrocarbons mixed
with nanochines that were nutritious, appetizing, and
pleasantly narcotic. Food and drink were not necessities,
but habits and they were enjoying a renewed popularity.
Like any credible Methuselah, Julian was often the model
on how best to do archaic oddities.

The woman lay on top of the bed. Her current body was
a hologram laid over her mechanical core. It was a traditional body, probably worn for his pleasure; no wings or
fins or even more bizarre adornments. As it happened, she
had selected a build and complexion not very different
from Julian's first wife. A coincidence? Or had she actually
done research, and she already knew the answers to her
prying questions?

"Sip," he advised, handing her the drink.

Their hands brushed against one another, shaped light
touching its equivalent. What each felt was a synthetically
generated sensation, basically human, intended to feel like
warm, water-filled skin.

The girl obeyed, smiling as she sipped, an audible slurp
amusing both of them.

"Here," she said, handing back the glass. "Your turn."

Julian glanced at the far wall. A universal window gave
them a live view of the Quick Shop, the image supplied by
one of the multitude of cameras hidden on the Buick's exterior. What held his interest was the old muscle car, a
Python with smoked glass windows. When he first saw
that car, three heads were visible. Now two of the heads
had gradually dropped out of sight, with the remaining

man still sitting in back, big eyes opened wide, making no attempt to hide his interest in the Buick's driver.

No one knew who the fat man was, or what he knew, much less what his intentions might be. His presence had been a complete surprise, and what he had done with those militia members, pulling them back as he did as well as the rest of it, had left the refugees more startled than grateful, and more scared than any time since leaving the Nest.

Julian had gone to that store with the intent of suffering a clumsy, even violent interrogation. A militia encounter was meant to give them authenticity. And more importantly, to give Julian experience—precious and sobering firsthand experience with the much-changed world around them.

A world that he hadn't visited for more than a millennium, Nesttime.

Since he last looked, nothing of substance had changed at that ugly store. And probably nothing would change for a long while. One lesson that no refugee needed, much less craved, was that when dealing with that other realm, nothing helped as much as patience.

Taking a long, slow sip of their drink, he looked back at the woman—twenty days old; a virtual child—and without a shred of patience, she said, "You were sick, weren't you? I heard someone saying that's why you agreed to be Transmutated . . . five hundred and twelve days ago . . ."

"No." He offered a shy smile. "And it wasn't because I wanted to live this way, either. To be honest, I've always been conservative. In that world, and this one, too."

She nodded amiably, waiting.

"It was my daughter," he explained. "She was sick. An incurable leukemia." Again he offered the shy smile, adding, "She was nine years old, and terrified. I could save her life by agreeing to her Transmutation, but I couldn't just abandon her to life in the Nest . . . making her into an orphan, basically . . ."

"I see," his lover whispered.

Then after a respectful silence, she asked, "Where's your daughter now?"

"Dead."

"Of course . . ." Not many people were lucky enough to live five hundred days in a Nest; despite shields, a single heavy nucleus could still find you, ravaging your mind, extinguishing your very delicate soul. "How long ago . . . did it happen . . . ?"

"This morning," he replied. "In the attack."

"Oh . . . I'm very sorry . . ."

With the illusion of shoulders, Julian shrugged. Then with his bittersweet voice, he admitted, "It already seems long ago."

Winemaster headed north into Iowa, then did the unexpected, making the sudden turn east when he reached the new Tollway.

Blaine shadowed him. He liked to keep two minutes between the Buick and his little Tokamak, using the FBI's recon network to help monitor the situation. But the network had been compromised in the past, probably more often than anyone knew, which meant that he had to occasionally pay the Tollway a little extra to boost his speed, the gap closing to less than fifteen seconds. Then with the optics in his windshield, he would get a good look at what might or might not be Julian Winemaster—a stiffly erect gentleman who kept one hand on the wheel, even when the AI-managed road was controlling every vehicle's speed and direction, and doing a better job of driving than any human could do.

Iowa was half-beautiful, half-bleak. Some fields looked tended, genetically tailored crops planted in fractal patterns and the occasional robot working carefully, pulling weeds and killing pests as it spider-walked back and forth.

But there were long stretches where the farms had been abandoned, wild grasses and the spawn of last year's crops coming up in ragged green masses. Entire neighborhoods had pulled up and gone elsewhere. How many farmers had accepted the Transmutation, in other countries or illegally? Probably only a fraction of them, Blaine knew. Habit-bound and suspicious by nature, they'd never agree to the dismantlement of their bodies, the transplantation of their crusty souls. No, what happened was that farms were simply falling out of production, particularly where the soil was marginal. Yields were still improving in a world where the old-style population was tumbling. If patterns held, most of the arable land would soon return to prairie and forest. And eventually, the entire human species wouldn't fill so much as one of these abandoned farms . . . leaving the old world entirely empty . . . if those patterns were allowed to hold, naturally . . .

Unlike Winemaster, Blaine kept neither hand on the wheel, trusting the AIs to look after him. He spent most of his time watching the news networks, keeping tabs on moods more than facts. What had happened in Kansas was still the big story. By noon, more than twenty groups and individuals had claimed responsibility for the attack. Officially, the Emergency Federal Council deplored any sense-less violence—a cliché which implied that sensible violence was an entirely different question. When asked about the government's response, the President's press sec-retary looked at the world with a stony face, saying, "We're investigating the regrettable incident. But the fact remains, it happened outside our borders. We are observers here. The Shawnee Nest was responsible for its own secu-rity, just as every other Nest is responsible . . ."

Questions came in a flurry. The press secretary pointed to a small, severe-looking man in the front row—a reporter for the Christian Promise organization. "Are we taking any precautions against counterattacks?" the reporter inquired.

Then, not waiting for an answer, he added, "There have been reports of activity in the other Nests, inside the United States and elsewhere."

A tense smile was the first reply.

Then the stony face told everyone, "The President and the Council have taken every appropriate precaution. As for any activity in any Nest, I can only say: We have everything perfectly well in hand."

"Is anything left of the Shawnee Nest?" asked a second reporter.

"No." The press secretary was neither sad nor pleased. "Initial evidence is that the entire facility has been sterilized."

A tenacious gray-haired woman—the perpetual symbol of the Canadian Newsweb—called out, "Mr. Secretary . . . Lennie—!"

"Yes, Cora . . ."

"How many were killed?"

"I wouldn't know how to answer that question, Cora . . ."

"Your government estimates an excess of one hundred million. If the entire Nest was sterilized, as you say, then we're talking about more than two-thirds of the current U.S. population."

"Legally," he replied, "we are talking about machines."

"Some of those machines were once your citizens," she mentioned.

The reporter from Christian Promise was standing nearby. He grimaced, then muttered bits of relevant Scripture.

"I don't think this is the time or place to debate what life is or isn't," said the press secretary, juggling things badly.

Cora persisted. "Are you aware of the Canadian position on this tragedy?"

"Like us, they're saddened."

"They've offered sanctuary to any survivors of the blast—"

"Except there are none," he replied, his face pink as granite.

"But if there were? Would you let them move to another Nest in the United States, or perhaps to Canada . . . ?"

There was a pause, brief and electric.

Then with a flat, cool voice, the press secretary reported, "The McGrugger Bill is very specific. Nests may exist only in sealed containment facilities, monitored at all times. And should any of the microchines escape, they will be treated as what they are . . . grave hazards to normal life . . . and this government will not let them roam at will . . . !"

Set inside an abandoned salt mine near Kansas City, the Shawnee Nest had been one of the most secure facilities of its kind ever built. Its power came from clean geothermal sources. Lead plates and intricate defense systems stood against natural hazards as well as more human threats. Thousands of government-loyal AIs, positioned in the surrounding salt, did nothing but watch its borders, making certain that none of the microchines could escape. That was why the thought that local terrorists could launch any attack was so ludicrous. To have that attack succeed was simply preposterous. Whoever was responsible for the bomb, it was done with the abeyance of the highest authorities. No sensible soul doubted it. That dirty little nuke had Federal fingerprints on it, and the attack was planned carefully, and its goals were instantly apparent to people large and small.

Julian had no doubts. He had enemies, vast and malicious, and nobody was more entitled to his paranoias.

Just short of Illinois, the Buick made a long-scheduled stop.

Julian took possession of his clone at the last moment.
The process was supposed to be routine—a simple matter
of slowing his thoughts a thousandfold, then integrating
them with his body—but there were always phantom pains
and a sick falling sensation. Becoming a bloated watery
bag wasn't the strangest part of it. After all, the Nest was
designed to mimic this kind of existence. What gnawed at
Julian was the gargantuan sense of Time: Half an hour in
this realm was nearly a month in his realm. No matter how
brief the stop, Julian would feel a little lost when he returned, a step behind the others, and far more emotionally
drained than he would ever admit.

By the time the car had stopped, Julian was in full control of the body. His body, he reminded himself. Climbing
out into the heat and brilliant sunshine, he felt a purposeful stiffness in his back and the familiar ache running down
his right leg. In his past life, he was plagued by sciatica
pains. It was one of many ailments that he hadn't missed
after his Transmutation. And it was just another detail that
someone had thought to include, forcing him to wince and
stretch, showing the watching world that he was their
flavor of mortal.

Suddenly another old pain began to call to Julian.

Hunger.

His duty was to fill the tank, then do everything expected of a road-weary driver. The rest area was surrounded by the Tollway, gas pumps surrounding a fast
food/playground complex. Built to handle tens of thousands of people daily, the facility had suffered with the
civil chaos, the militias and the plummeting populations.
A few dozen travelers went about their business in near-
solitude, and presumably a team of state or Federal agents
were lurking nearby, using sensors to scan for those who
weren't what they seemed to be.

Without incident, Julian managed the first part of his
mission. Then he drove a tiny distance and parked, repeat-

ing his stiff climb out of the car, entering the restaurant and steering straight for the rest room.

He was alone, thankfully.

The diagnostic urinal gently warned him to drink more fluids, then wished him a lovely day.

Taking the advice to heart, Julian ordered a bucket-sized iced tea along with a cultured guinea hen sandwich.

"For here or to go?" asked the automated clerk.

"I'm staying," he replied, believing it would look best.

"Thank you, sir. Have a lovely day."

Julian sat in the back booth, eating slowly and mannerly, scanning the pages of someone's forgotten e-paper. He made a point of lingering over the trite and trivial, concentrating on the comics with their humanized cats and cartoonish people, everyone playing out the same jokes that must have amused him in the very remote past.

"How's it going?"

The voice was slow and wet. Julian blanked the page, looking over his shoulder, betraying nothing as his eyes settled on the familiar wide face. "Fine," he replied, his own voice polite but distant. "Thank you."

"Is it me? Or is it just too damned hot to live out there . . . ?"

"It is hot," Julian conceded.

"Particularly for the likes of me." The man settled onto a plastic chair bolted into the floor with clown heads. His lunch buried his little table: three sandwiches, a greasy sack of fried cucumbers, and a tall chocolate shake. "It's murder when you're fat. Let me tell you . . . I've got to be careful in this weather. I don't move fast. I talk softly. I even have to ration my thinking. I mean it! Too many thoughts, and I break out in a killing sweat!"

Julian had prepared for this moment. Yet nothing was happening quite like he or anyone else had expected.

Saying nothing, Julian took a shy bite out of his sandwich.

"You look like a smart guy," said his companion. "Tell me. If the world's getting emptier, like everyone says, why am I still getting poorer?"

"Excuse me?"

"That's the way it feels, at least." The man was truly fat, his face smooth and youthful, every feature pressed outward by the remnants of countless lunches. "You'd think that with all the smart ones leaving for the Nests . . . you'd think guys like you and me would do pretty well for ourselves. You know?"

Using every resource, the refugees had found three identities for this man: He was a salesman from St. Joseph, Missouri. Or he was a Federal agent working for the Department of Technology, in its Enforcement division, and his salesman identity was a cover. Or he was a charter member of the Christian Promise organization, using that group's political connections to accomplish its murderous goals.

What does he want? Julian asked himself.

He took another shy bite, wiped his mouth with a napkin, then offered his own question. "Why do you say that . . . that it's the smart people who are leaving . . . ?"

"That's what studies show," said a booming, unashamed voice. "Half our people are gone, but we've lost ninety percent of our scientists. Eighty percent of our doctors. And almost every last member of Mensa . . . which between you and me is a good thing, I think . . . !"

Another bite, and wipe. Then with a genuine firmness, Julian told him, "I don't think we should be talking. We don't know each other."

A huge cackling laugh ended with an abrupt statement:

"That's why we should talk. We're strangers, so where's the harm?"

Suddenly the guinea hen sandwich appeared huge and inedible. Julian set it down and took a gulp of tea.

His companion watched him, apparently captivated.

Julian swallowed, then asked, "What do you do for a living?"

"What I'm good at." He unwrapped a hamburger, then took an enormous bite, leaving a crescent-shaped sandwich and a fine glistening stain around his smile. "Put it this way, Mr. Winemaster. I'm like anyone. I do what I hope is best."

"How do you—?"

"Your name? The same way I know your address, and your social registration number, and your bank balance, too." He took a moment to consume half of the remaining crescent, then while chewing, he choked out the words, "Blaine. My name is. If you'd like to use it."

Each of the man's possible identities used Blaine, either as a first or last name.

Julian wrapped the rest of his sandwich in its insulated paper, watching his hands begin to tremble. He had a pianist's hands in his first life but absolutely no talent for music. When he went through the Transmutation, he'd asked for a better ear and more coordination—both of which were given to him with minimal fuss. Yet he'd never learned how to play, not even after five hundred days. It suddenly seemed like a tragic waste of talent, and with a secret voice, he promised himself to take lessons, starting immediately.

"So, Mr. Winemaster . . . where are you heading . . . ?"

Julian managed another sip of tea, grimacing at the bitter taste.

"Someplace east, judging by what I can see . . ."

"Yes," he allowed. Then he added, "Which is none of your business."

Blaine gave a hearty laugh, shoving the last of the burger deep into his gaping mouth. Then he spoke, showing off the masticated meat and tomatoes, telling his new friend, "Maybe you'll need help somewhere up ahead. Just maybe. And if that happens, I want you to think of me."

"You'll help me, will you?"

The food-stuffed grin was practically radiant. "Think of me," he repeated happily. "That's all I'm saying."

For a long while, the refugees spoke and dreamed of nothing but the mysterious Blaine. Which side did he represent? Should they trust him? Or move against him? And if they tried to stop the man, which way was best? Sabotage his car? Drug his next meal? Or would they have to do something genuinely horrible?

But there were no answers, much less a consensus. Blaine continued shadowing them, at a respectful distance; nothing substantial was learned about him; and despite the enormous stakes, the refugees found themselves gradually drifting back into the moment-by-moment business of ordinary life.

Couples and amalgamations of couples were beginning to make babies.

There was a logic: Refugees were dying every few minutes, usually from radiation exposure. The losses weren't critical, but when they reached their new home—the deep cold rock of the Canadian Shield—they would need numbers, a real demographic momentum. And logic always dances with emotion. Babies served as a tonic to the adults. They didn't demand too many resources, and they forced their parents to focus on more manageable problems, like building tiny bodies and caring for needy souls.

Even Julian was swayed by fashion.

With one of his oldest women friends, he found himself hovering over a crystalline womb, watching nanochines sculpt their son out of single atoms and tiny electric breaths.

It was only Julian's second child.

As long as his daughter had been alive, he hadn't seen the point in having another. The truth was that it had always disgusted him to know that the children in the Nest were manufactured—there was no other word for it—and he didn't relish being reminded that he was nothing, more or less, than a fancy machine among millions of similar machines.

Julian often dreamed of his dead daughter. Usually she was on board their strange ark, and he would find a note from her, and a cabin number, and he would wake up smiling, feeling certain that he would find her today. Then he would suddenly remember the bomb, and he would start to cry, suffering through the wrenching, damning loss all over again.

Which was ironic, in a fashion.

During the last nineteen months, father and daughter had gradually and inexorably drifted apart. She was very much a child when they came to the Nest, as flexible as her father wasn't, and how many times had Julian lain awake in bed, wondering why he had ever bothered being Transmutated. His daughter didn't need him, plainly. He could have remained behind. Which always led to the same questions: When he was a normal human being, was he genuinely happy? Or was his daughter's illness simply an excuse . . . a spicy bit of good fortune that offered an escape route . . . ?

When the Nest was destroyed, Julian survived only through more good fortune. He was as far from the epicenter as possible, shielded by the Nest's interior walls and emergency barricades. Yet even then, most of the people near him were killed, an invisible neutron rain scrambling their minds. That same rain had knocked him unconscious just before the firestorm arrived, and if an autodoc hadn't found his limp body, then dragged him into a shelter, he would have been cremated. And of course if the Nest hadn't devised its elaborate escape plan, stockpiling the Buick and cloning equipment outside the Nest, Julian would have had no choice but to remain in the rubble, fighting to survive the next moment, and the next.

But those coincidences happened, making his present life feel like the culmination of some glorious Fate.

The secret truth was that Julian relished his new importance, and he enjoyed the pressures that came with each bathroom break and every stop for gas. If he died now, between missions, others could take his place, leading Winemaster's cloned body through the needed motions . . . but they wouldn't do as well, Julian could tell himself . . . a secret part of him wishing that this bizarre, slow-motion chase would never come to an end . . .

The Buick stayed on the Tollway through northern Illinois, slipping beneath Chicago before skipping across a sliver of Indiana. Julian was integrated with his larger self several times, going through the motions of the stiff, tired, and hungry traveler. Blaine always arrived several minutes later, never approaching his quarry, always finding gas at different pumps, standing outside the rest rooms, waiting to show Julian a big smile but never uttering so much as a word in passing.

A little after midnight, the Buick's driver took his hand off the wheel, lay back, and fell asleep. Trusting the Tollway's driving was out of character, but with Blaine trailing them and the border approaching, no one was eager to waste time in a motel bed.

At two in the morning, Julian was also asleep, dipping in and out of dreams. Suddenly a hand took him by the shoulder, shaking him, and several voices, urgent and close, said, "We need you, Julian. Now."

In his dreams, a thousand admiring faces were saying, "We need you."

Julian awoke.

His cabin was full of people. His mate had been ushered

away, but his unborn child, nearly complete now, floated in his bubble of blackened crystal, oblivious to the nervous air and the tight, crisp voices.

"What's wrong?" Julian asked.

"Everything," they assured.

His universal window showed a live feed from a security camera on the North Dakota–Manitoba border. Department of Technology investigators, backed up by a platoon of heavily armed Marines, were dismantling a Toyota Sunrise. Even at those syrupy speeds, the lasers moved quickly, leaving the vehicle in tiny pieces that were photographed, analyzed, then fed into a state-of-the-art decontamination unit.

"What is this?" Julian sputtered.

But he already knew the answer.

"There was a second group of refugees," said the President, kneeling beside his bed. She was wearing an oversized face—a common fashion, of late—and with a very calm, very grim voice, she admitted, "We weren't the only survivors."

They had kept it a secret, at least from Julian. Which was perfectly reasonable, he reminded himself. What if he had been captured? Under torture, he could have doomed that second lifeboat, and everyone inside it . . .

"Is my daughter there?" he blurted, uncertain what to hope for.

The President shook her head. "No, Julian."

Yet if two arks existed, couldn't there be a third? And wouldn't the President keep its existence secret from him, too?

"We've been monitoring events," she continued. "It's tragic, what's happening to our friends . . . but we'll be able to adjust our methods . . . for when we cross the border . . ."

He looked at the other oversized faces. "But why do you need me? We won't reach Detroit for hours."

The President looked over her shoulder. "Play the recording."

Suddenly Julian was looking back in time. He saw the Sunrise pull up to the border post, waiting in line to be searched. A pickup truck with Wyoming plates pulled up behind it, and out stepped a preposterously tall man brandishing a badge and a handgun. With an eerie sense of purpose, he strode up to the little car, took aim, and fired his full clip through the driver's window. The body behind the wheel jerked and kicked as it was ripped apart. Then the murderer reached in and pulled the corpse out through the shattered glass, shouting at the Tech investigators.

"I've got them! Here! For Christ's sake, help me!"

The image dissolved, the window returning to the real-time, real-speed scene.

To himself, Julian whispered, "No, it can't be . . ."

The President took his hands in hers, their warmth a comfortable fiction. "We would have shown you this as it was happening, but we weren't sure what it meant."

"But you're sure now?"

"That man followed our people. All the way from Nebraska." She shook her head, admitting, "We don't know everything, no. For security reasons, we rarely spoke with those other survivors—"

"What are we going to do?" Julian growled.

"The only reasonable thing left for us." She smiled in a sad fashion, then warned him, "We're pulling off the Tollway now. You still have a little while to get ready . . ."

He closed his eyes, saying nothing.

"Not as long as you'd like, I'm sure . . . but with this sort of thing, maybe it's best to hurry . . ."

There were no gas pumps or restaurants in the rest area. A small divided parking lot was surrounded by trees and fake log cabin lavatories that in turn were sandwiched between broad lanes of moonlit pavement. The parking lot was empty. The only traffic was a single truck in the westbound freighter lane, half a dozen trailers towed along in its wake. Julian watched the truck pass, then walked into the darkest shadows, and kneeled.

The security cameras were being fed false images—images that were hopefully more convincing than the ludicrous log cabins. Yet even when he knew that he was safe, Julian felt exposed. Vulnerable. The feeling worsened by the moment, becoming a black dread, and by the time the Tokamak pulled to stop, his newborn heart was racing, and his quick damp breath tasted foul.

Blaine parked two slots away from the sleeping Buick. He didn't bother looking through the windows. Instead, guided by intuition or hidden sensor, he strolled toward the men's room, hesitated, then took a few half-steps toward Julian, passing into a patch of moonlight.

Using both hands, Julian lifted his weapon, letting it aim itself at the smooth, broad forehead.

"Well," said Blaine, "I see you're thinking about me."

"What do you want?" Julian whispered. Then with a certain clumsiness, he added, "With me."

The man remained silent for a moment, a smile building.

"Who am I?" he asked suddenly. "Ideas? Do you have any?"

Julian gulped a breath, then said, "You work for the government." His voice was testy, pained. "And I don't know why you're following me!"

Blaine didn't offer answers. Instead he warned his audience, "The border is a lot harder to pierce than you think."

"Is it?"

"Humans aren't fools," Blaine reminded him. "After all, they designed the technologies used by the Nests, and they've had just as long as you to improve on old tricks."

"People in the world are getting dumber," said Julian. "You told me that."

"And those same people are very scared, very focused," his opponent countered. "Their borders are a priority to them. You are their top priority. And even if your thought processes are accelerated a thousandfold, they've got AIs who can blister you in any race of intellect. At least for the time being, they can."

Shoot him, an inner voice urged.

Yet Julian did nothing, waiting silently, hoping to be saved from this onerous chore.

"You can't cross into Canada without me," Blaine told him.

"I know what happened . . ." Julian felt the gun's barrel adjusting itself as his hands grew tired and dropped slightly. "Up in North Dakota . . . we know all about it . . ."

It was Blaine's turn to keep silent.

Again, Julian asked, "Who are you? Just tell me that much."

"You haven't guessed it, have you?" The round face seemed genuinely disappointed. "Not even in your wildest dreams . . ."

"And why help us?" Julian muttered, saying too much.

"Because in the long run, helping you helps me."

"How?"

Silence.

"We don't have any wealth," Julian roared. "Our homes were destroyed. By you, for all I know—"

The man laughed loudly, smirking as he began to turn away. "You've got some time left. Think about the possibilities, and we'll talk again."

Julian tugged on the trigger. Just once.

Eighteen shells pierced the back of Blaine's head, then worked down the wide back, devastating every organ even as the lifeless body crumpled. Even a huge man falls fast,

Julian observed. Then he rose, walking on weak legs, and with his own aim, he emptied the rest of his clip into the gore.

It was easy, pumping in those final shots.

What's more, shooting the dead carried an odd, unexpected satisfaction—which was probably the same satisfaction that the terrorists had felt when their tiny bomb destroyed a hundred million soulless machines.

With every refugee watching, Julian cut open the womb with laser shears.

Julian Jr. was born a few seconds after two-thirty A.M., and the audience, desperate for a good celebration, nearly buried the baby with gifts and sweet words. Yet nobody could spoil him like his father could. For the next few hours, Julian pestered his first son with love and praise, working with a manic energy to fill every need, every whim. And his quest to be a perfect father only grew worse. The sun was beginning to show itself; Canada was waiting over the horizon; but Julian was oblivious, hunched over the toddler with sparkling toys in both hands, his never-pretty voice trying to sing a child's song, nothing half as important in this world as making his son giggle and smile . . . !

They weren't getting past the border. Their enemies were too clever, and too paranoid. Julian could smell the inevitable, but because he didn't know what else to do, he went through the motions of smiling for the President and the public, saying the usual brave words whenever it was demanded of him.

Sometimes Julian took his boy for long rides around the lifeboat.

During one journey, a woman knelt and happily teased the baby, then looked up at the famous man, mentioning in

an offhanded way, "We'll get to our new home just in time
for him to grow into it."

Those words gnawed at Julian, although he was powerless to explain why.

By then the sun had risen, its brilliant light sweeping
across a sleepy border town. Instead of crossing at Detroit,
the refugees had abandoned the Tollway, taking an old
highway north to Port Huron. It would be easier here, was
the logic. The prayer. Gazing out the universal window, Julian looked at the boarded-up homes and abandoned busi-
nesses, cars parked and forgotten, weeds growing in every
yard, every crack. The border cities had lost most of their
people in the last year-plus, he recalled. It was too easy and
too accepted, this business of crossing into a land where it
was still legal to be remade. In another year, most of the
United States would look this way, unless the government
took more drastic measures such as closing its borders, or
worse, invading its wrongminded neighbors . . . !

Julian felt a deep chill, shuddering.

That's when he suddenly understood. Everything. And
in the next few seconds, after much thought, he knew precisely what he had to do.

Assuming there was still time . . .

A dozen cars were lined up in front of the customs station.
The Buick had slipped in behind a couple on a motorcycle.
Only one examination station was open, and every traveler
was required to first declare his intentions, then permanently give up his citizenship. It would be a long wait. The
driver turned the engine off, watching the Marines and
Tech officials at work, everything about them relentlessly
professional. Three more cars pulled up behind him, including a Tokamak, and he happened to glance at the
rearview screen when Blaine climbed out, walking with a
genuine bounce, approaching on the right and rapping on

the passenger window with one fat knuckle, then stooping down and smiling through the glass, proving that he had made a remarkable recovery since being murdered.

Julian unlocked the door for him.

With a heavy grunt, Blaine pulled himself in and shut the door, then gave his companion a quick wink.

Julian wasn't surprised. If anything, he was relieved, telling his companion, "I think I know what you are."

"Good," said Blaine. "And what do your friends think?"

"I don't know. I never told them." Julian took the steering wheel in both hands. "I was afraid that if I did, they wouldn't believe me. They'd think I was crazy, and dangerous. And they wouldn't let me come here."

The line was moving, jerking forward one car-length. Julian started the Buick and crept forward, then turned it off again.

With a genuine fondness, Blaine touched him on a shoulder, commenting, "Your friends might pull you back into their world now. Have you thought of that?"

"Sure," said Julian. "But for the next few seconds, they'll be too confused to make any big decisions."

Lake Huron lay on Blaine's left, vast and deeply blue, and he studied the picket boats that dotted the water, bristling with lasers that did nothing but flip back and forth, back and forth, incinerating any flying object that appeared even remotely suspicious.

"So tell me," he asked his companion, "why do you think I'm here?"

Julian turned his body, the cultured leather squeaking beneath him. Gesturing at Port Huron, he said, "If these trends continue, everything's going to look that way very soon. Empty. Abandoned. Humans will have almost vanished from this world, which means that perhaps someone else could move in without too much trouble. They'll find houses, and good roads to drive on, and a communication system already in place. Ready-made lives, and practically free for the taking."

"What sort of someone?"

"That's what suddenly occurred to me." Julian took a deep breath, then said, "Humans are making themselves smaller, and faster. But what if something other than humans is doing the same thing? What if there's something in the universe that's huge, and very slow by human standards, but intelligent nonetheless. Maybe it lives in cold places between the stars. Maybe somewhere else. The point is, this other species is undergoing a similar kind of transformation. It's making itself a thousand times smaller, and a thousand times quicker, which puts it roughly equal to this." The frail face was smiling, and he lifted his hands from the wheel. "Flesh and blood, and bone . . . these are the high-technology materials that build your version of microchines!"

Blaine winked again, saying, "You're probably right. If you'd explained it that way, your little friends would have labeled you insane."

"But am I right?"

There was no reason to answer him directly. "What about me, Mr. Winemaster? How do you look at me?"

"You want to help us." Julian suddenly winced, then shuddered. But he didn't mention it, saying, "I assume that you have different abilities than we do . . . that you can get us past their sensors—"

"Is something wrong, Mr. Winemaster?"

"My friends . . . they're trying to take control of this body . . ."

"Can you deal with them?"

"For another minute. I changed all the control codes." Again, he winced. "You don't want the government aware of you, right? And you're trying to help steer us and then away from war . . . during this period of transition—"

"The way we see it," Blaine confessed, "the chance of

a worldwide cataclysm is just about one in three, and worsening."

Julian nodded, his face contorting in agony. "If I accept your help . . . ?"

"Then I'll need yours." He set a broad hand on Julian's neck. "You've done a remarkable job hiding yourselves. You and your friends are in this car, but my tools can't tell me where. Not without more time, at least. And that's time we don't have . . ."

Julian stiffened, his clothes instantly soaked with perspiration.

Quietly, quickly, he said, "But if you're really a government agent . . . here to fool me into telling you . . . everything . . . ?"

"I'm not," Blaine promised.

A second examination station had just opened; people were maneuvering for position, leaving a gap in front of them.

Julian started his car, pulling forward. "If I do tell you . . . where we are . . . they'll think that I've betrayed them . . . !"

The Buick's anticollision system engaged, bringing them to an abrupt stop.

"Listen," said Blaine. "You've got only a few seconds to decide—"

"I know . . ."

"Where, Mr. Winemaster? Where?"

"Julian," he said, wincing again.

"Julian."

A glint of pride showed in the eyes. "We're not . . . in the car . . ." Then the eyes grew enormous, and Julian tried shouting the answer . . . his mind suddenly losing its grip on that tiny, lovely mouth . . .

Blaine swung with his right fist, shattering a cheekbone with his first blow, killing the body before the last blow.

By the time the Marines had surrounded the car, its interior was painted with gore, and in horror, the soldiers watched the madman—he couldn't be anything but insane—calmly rolled down his window and smiled with a blood-rimmed mouth, telling his audience, "I had to kill him. He's Satan."

A hardened lieutenant looked in at the victim, torn open like a sack, and she shivered, moaning aloud for the poor man.

With perfect calm, Blaine declared, "I had to eat his heart. That's how you kill Satan. Don't you know?"

For disobeying orders, the President declared Julian a traitor, and she oversaw his trial and conviction. The entire process took less than a minute. His quarters were remodeled to serve as his prison cell. In the next ten minutes, three separate attempts were made on his life. Not everyone agreed with the court's sentence, it seemed. Which was understandable. Contact with the outside world had been lost the instant Winemaster died. The refugees and their lifeboat were lost in every kind of darkness. At any moment, the Tech specialists would throw them into a decontamination unit and they would evaporate without warning. And all because they'd entrusted themselves to an old DNA-born human who never really wanted to be Transmutated in the first place, according to at least one of his former lovers . . .

Ostensibly for security reasons, Julian wasn't permitted visitors.

Not even his young son could be brought to him, nor was he allowed to see so much as a picture of the boy.

Julian spent his waking moments pacing back and forth in the dim light, trying to exhaust himself, then falling into a hard sleep, too tired to dream at all, if he was lucky . . .

Before the first hour was finished, he had lost all track of time.

After nine full days of relentless isolation, the universe had shriveled until nothing existed but his cell, and him, his memories indistinguishable from fantasies.

On the tenth day, the cell door opened.

A young man stepped in, and with a stranger's voice, he said, "Father."

"Who are you?" asked Julian.

His son didn't answer, giving him the urgent news instead. "Mr. Blaine finally made contact with us, explaining what he is and what's happened so far, and what will happen . . . !"

Confusion wrestled with a fledgling sense of relief.

"He's from between the stars, just like you guessed, Father. And he's been found insane for your murder. Though of course you're not dead. But the government believes there was a Julian Winemaster, and it's holding Blaine in a Detroit hospital, and he's holding us. His metabolism is augmenting our energy production, and when nobody's watching, he'll connect us with the outside world."

Julian couldn't imagine such a wild story: It had to be true!

"When the world is safe, in a year or two, he'll act cured or he'll escape—whatever is necessary—and he'll carry us wherever we want to go."

The old man sat on his bed, suddenly exhausted.

"Where would you like to go, Father?"

"Out that door," Julian managed. Then a wondrous thought took him by surprise, and he grinned, saying, "No. I want to be like Blaine was. I want to live between the stars, to be huge and cold, and slow . . .

"Not today, maybe . . .

"But soon . . . definitely soon . . . !"

MORE ADVENTURES
ON OTHER PLANETS

Michael Cassutt

Sometimes the problem with going beyond flesh is that even going millions of miles away, to the frozen surface of a hostile alien world, is not going nearly far enough . . .

As a print author, Michael Cassutt is mostly known for his incisive short work, but he has worked intensively in the television industry over the past few decades, where he is a major mover and shaker. He was co-executive producer for Showtime's The Outer Limits—*which won a CableAce Award for best dramatic series—and also served in the same or similar capacities for series such as* Eerie, Indiana *and* StrangeLuck, *as well as having worked as the story editor for* Max Headroom, *as a staff writer on* The Twilight Zone, *and having contributed scripts to* Farscape, Stargate SG-1, *and many other television series. He also contributes a regular column on science fiction in films and television to* Science Fiction Weekly. *His books include the novels* Star Country, Dragon Season, *and* Missing Man, *the anthology* Sacred Visions, *co-edited with Andrew M. Greeley, and a biographical encyclopedia,* Who's Who in Space: the First 25 Years. *He also collaborated with the late astronaut Deke Slayton on Slayton's autobiography,* Deke!. *His most recent book is the historical thriller* Red Moon.

This is what they used to call a cute meet, back when movies were made by people like Ernst Lubitsch or Billy

Wilder, when movies had plots and dialogue, when life and love had rules, back in the last century. A handsome officer in the Soviet embassy (does that tell you how long ago?) picks up the phone one day and hears a lilting female voice asking him if he can tell her, please, what is Lenin's middle name. "It's for my crossword puzzle."

Affronted, the officer snaps, "To dignify that question would be an insult to the Soviet Union!" And slams down the phone.

But not before he hears a lovely laugh.

That evening the officer goes to the British Embassy for some reception, and hears that same laugh emerging from the oh so luscious mouth of an English woman who should probably be Audrey Hepburn. Smitten, the officer walks up to Miss Hepburn, bows, and says, "Ilyich."

And so the story begins.

And so *our* story begins. Only—

Look, you're going to have to be patient with me. Because the couple is not just a couple. It's more of a quartet. And two of the individuals aren't even *people*.

Picture the surface of Europa, the icy moon of Jupiter. It is midday, local time, but the sky is black: What little atmosphere Europa possesses is insufficient to scatter enough light to give it a color. The combination of ice, snow, and rock create a patchwork of white and gray, something like a chessboard with no straight lines.

Europa is tectonically active, about ten times as bouncy as anyplace on earth, so the landscape is marked by jagged upthrusts and creepy fissures known as cycloids.

But forget the landscape and the color of the sky. What really catches your attention is the striped ball that is Jupiter, looming overhead like a gigantic jack-o'-lantern. It actually seems to press down on the snowy landscape. What makes it a little worse is that since Europa is tide-locked, always keeping the same face toward its giant mother, if you happen to be working on that side of Europa, Jupiter is always there!

And so are several elements of the J^2E^2, the Joint Jupiter-Europan Expedition, three tiny rovers that have been operating on the icy plains for two years, scouting the site for the "permanent" Hoppa Station and erecting such necessary equipment as a shelter (even machines get cold on Europa), a radiothermal power plant, and the communications array.

On this particular day, rover element one, also known as "Earl," is approximately seven kilometers north of Hoppa when he receives a query from a source in motion (his comm gear is sophisticated enough to detect a slight Doppler effect) for range-rate data.

Element Earl can't see the source: His visual sensor is a hardy multispectral charged-couple device that is excellent for showing a view forward and all around. It lacks, however, a tilt mechanism that will let it see *up*.

Nor, given the priorities in his guidance system, can he presently provide range-rate data. In the burst of bits that made up rover-speak, Element Earl says, more or less, "I'm a Pathfinder-class rover element. You should be talking to the base unit at Hoppa Station."

He would think no more about the contact, except that there is a message of sorts embedded in the acknowledgment that suggests . . . compatibility. More than seems to exist between the Dopplering radio source and the base unit at Hoppa, in any case.

The Dopplering source is, in fact, a series of follow-up J^2E^2 packages designed to conduct the search for life in the dark, frigid ocean under Europa's icy crust.

All of these elements are wrapped inside a landing bag dropped from a mission bus launched from Earth two years after the initial bunch that included Element Earl and propelled Europa-ward by lightsail. The bus has burned into orbit around Europa, then waited for a command from La Jolla to separate the bag and its retro system.

The follow-up flight has been marred by software glitches, some of them due to undetected programming lapses back in the avionics lab in La Jolla, others to the assault of Jupiter's magnetic field. After all, the chips are only hardened against electromagnetic pulse from a nuclear weapon, not the steady and relentless assault of charged Jovian particles. Like a human trained to withstand a stomach punch only to find himself dragged behind a truck, the bus has suffered some damage.

Which is why one of its four elements, soon to be known as "Rebecca," goes on-line during the descent phase as a backup to the lander's systems, which are having a tough time locking on to the signal from Hoppa Station. Not to prolong the suspense, the landing package arrives safely, bouncing half a dozen times on the icy plain, punching holes in itself by design, and eventually disgorging four new elements.

It is only a week later when Element Earl, returning to station for thermal reasons, happens to detect (not *see*: His visual sensor is usually turned off to conserve power and he was simply retracing his original route) four new arrivals—the drilling, cargo, submersible, and portable power rover elements that will soon begin the search for life.

He passes close enough to the drilling rover, which is currently deploying its array, since diagnostics show it to have been damaged in the rolling, rocking landing. It so happens that the array wasn't damaged. But in the stream of bits flowing from the drilling rover to the Hoppa central unit and splashing from one rover to another, Element Earl notes the familiar signature of Element Rebecca.

As a bit of a joke, he aims his dish at hers, and feeds her the range-rate data she had asked for earlier.

•   •   •

Mission control for J^2E^2 is in a crumbling three-story structure in the bad part of La Jolla, south of the Cove and bordering on the aptly-named Mission Beach. The building formerly housed an Internet service provider. The ISP had purchased and remodeled the place in 1998, hoping for business from the San Diego and North County high-tech communities, which were then wallowing in an unprecedented economic boom.

And did so for the better part of a decade, until a series of mergers closed the node. Then the AGC Corporation, newly formed by three researchers from UC-SD, just over the hill in La Jolla proper, leased the building for tests for their first real-time Superluminal Light Pulse Propogation/Emulation Regime (usually known as SLIPPER) on the 2012 asteroid Neva flyby. What the hell: The facility was already wired for fiber-optic and extreme bandwith, and was configured for electrical and thermal support of AGC's ten-petaflop computer.

That was eighteen years and five interplanetary missions ago, and while the guts of what is now the J^2E^2 mission control have continued to evolve, the exterior has been left alone. Which presents the staff with a problem. The ISP operation had never employed more than a dozen people, while the AGC SLIPPER project has thirty or more in the building at all times.

The parking lot is simply inadequate, and with public transport in this part of La Jolla (remember, this is California) limited to the occasional bus, with working hours staggered, with rents and home prices in La Jolla among the highest in the country . . . well, disputes are inevitable.

Earl Tolan pulls his battered Chevy pickup into the gated lot and drives up to space eleven, only to find a brand-new Volvo already there.

Tolan is fifty-nine, a senior operator on the J^2E^2 project after moving to AGC from Lockheed Martin, where he led teams through good times and bad for twenty years. He is not one to lose his temper without reason.

But today he happens to be returning to work after what should have been a quick visit to the doctor, a checkup which wound up taking four hours and has left him in a bad mood. So the site of this impudent little Volvo taking up his space launches him into a state of only theoretically controlled fury.

He squeals the truck around so that its tailgate backs up to the Volvo. This is a bit of a trick, given the confined space. Tolan has to drive up and over a curb and sidewalk median just to get into position.

Once on station, as ops guys are fond of saying, he drops the tailgate, hauls out a length of chain and a hook he usually uses for attaching the smaller of his two boats to a trailer, wedges the hook in the Volvo's rear bumper, and loops the chain around his trailer hitch.

Then he gets into the truck, puts it in low, and hauls the Volvo out of his space, a maneuver which takes him up and onto the sidewalk and into the driveway beyond. The Volvo, its gear in park and its brake set, makes a screeching sound with its tires, followed by an ominous undercarriage scraping, before fetching up onto the sidewalk median.

Where Tolan leaves it.

Wallowing in momentary self-satisfaction, he pulls around into his space. He is still quite angry, in fact, when he emerges from the truck and heads for the building entrance, where he brushes shoulders with a woman going the other way.

Had his mood been anything less than ultraviolet anger and disgust, Tolan would certainly have managed to side-step the charging woman while simultaneously noting her looks. Which, allowing for a certain air of growing confusion, are barely worth noting: She is a little over five feet, but adding stature with heeled sandals. A pair of gray slacks suggest muscular legs, and a vest worn over a J^2E^2 polo shirt does nothing to conceal the solidity within. Her hair is shoulder-length, dark, with a few lighter streaks, appropriate to her age, which is fiftyish. He thinks the eyes are green, but needs a closer look.

Not that he's inclined to give one. Twice-divorced, his sexual relationships are generally with women who would register as more attractive than this one on any visual scale.

What actually gets Tolan's attention is this woman's voice, which has what used to be called (in the days when people still consumed both) a whiskey and cigarette tone, tinged with some kind of Euro accent. Or perhaps it is the words she uses: "I'm gonna kill the son of a bitch who did this." Meaning haul her Volvo onto the median.

The woman calmly walks up to the vehicle, which still quivers in the aftermath of its relocation. She folds her arms, smiles with what could have been a touch of amusement.

Tolan can still make a clean escape, though he knows it won't be long before someone connects the evidentiary dots between Tolan's parking space, the skid marks from it to the Volvo's resting place. Besides, he is curious about the color of those eyes—so curious, he forgets his anger over the momentary theft of a parking place, and his frustration over two hours of unwarranted medical tests.

"I'm the son of a bitch," Tolan said.

She looks at him. Yes, green, with a charming set of smile lines. "Aren't you old enough to know better?"

This strikes Tolan as unfair, given that he is staring at sixty on his next birthday and has just had a medical experience all too appropriate for that age. "Apparently not."

To her great credit, she laughs. "I assume this was your space." He nods. "Well, I'm so new, I don't have an assigned one. And the guard did tell me you weren't likely to return today."

"Surprises all around." He holds out his hand. "Earl Tolan."

"Rebecca Marceau."

"I think we've met before."

"Cologne?" she says, then realizes where. She blushes. "Oh! Hoppa Station." Operators like Earl and Rebecca are often brought into the program without prior introductions. After all, they are usually mature professionals.

"Actually, about twelve klicks away," Earl says, wondering why he feels the need to be so precise.

You have to forget everything you think you know about space flight. The SLIPPER operators aren't astronauts. In fact, there are damned few astronauts here in 2026, just a few poor souls stuck going round and round the earth for months at a time in the crumbling EarthStar space station, hoping their work will somehow overcome the bone loss or radiation exposure or even psychological barriers that prevented a manned mission to Mars, not to mention even more distant locales such as Europa.

But exploration of the solar system continues, using unmanned vehicles which can be controlled from distances of tens of millions of miles, more or less in real-time, by human beings. The advantages are many: The vehicles can be smaller, they need only be built for a one-way trip, and using SLIPPER-linked human operators allows spacecraft builders to skip the lengthy and unpredictable development of artificial intelligence systems.

J^2E^2's mission control in La Jolla, then, is more like a virtual reality game den than a Shuttle-era firing room. Yes, there are the basic trajectory and electrical support stations, complete with consoles, and there is a big screen that displays telemetry from all of the many separate elements, along with selected camera views.

But the real work is done in the eight booths at the back

of the control room, where each operator strips naked and dons a skintight SLIPPER suit and helmet not awfully different from scuba gear, allowing her to link up in real-time with her avatar on Europa.

To see Jupiter looming permanently on the horizon.

To feel the shudders of the hourly quakes.

To hear the crunch of treads on ice.

To smell metal and composite baked by radiation.

You can even taste the surge of energy when linked to the generator for recharging.

It's all faux reality, of course, the work of clever programmers who have created a system which translates digital data from the elements themselves into simulated "feelings," then reverses the process, translating an operator's muscular impulse to reach, for example, into a command to rotate an antenna.

The best operators are those who know spacecraft and their limitations, who have proven that they can commit to a mission plan. People who simply like machines also make good operators. For J^2E^2, AGC tries to find those who can fit both matrixes.

And who are willing to take the risk of permanent nerve damage caused by the interface.

Rebecca operates Earl's truck as he rocks the Volvo. He has chained the two vehicles together, and is learning that undoing his prank is easier than doing it, since the tightness of the driveway is forcing Rebecca and the truck to pull the Volvo at an angle.

But she expertly guns the motor just as Earl gets the Volvo's front wheels on the pavement. With a *hump!* and a *whoof!* and a reasonable amount of scraping, the Volvo shoots free. "That was suspiciously close to good sex," Rebecca says, delicately wiping sweat from her eyes.

Now it is Earl's turn to blush, something he can't remember happening in years. (He is old enough to know better about this, too.) He had been thinking the same thing. "You like cars," he says, lamely, fitting her neatly into that subset of the operator personality matrix, something the operators do both consciously and instinctively, like long-lost tribesmen smelling each other.

"Guilty, Officer," she says, and looks at the truck, with its complement of nautical equipment. "And for you it must be boats."

"Two of them. A runabout and a forty-five-footer." The tribal recognition isn't strong enough to overcome their mutual antagonism. Note that there is no invitation to take a sail.

"See you on Europa."

On Europa, science is marching more slowly than usual. Element Rebecca is tasked with drilling a hole through the icy crust at a site seven kilometers north of Hoppa Station. The same spot Element Earl was scouting the day the science package arrived.

Now, from a distance, at the macro level, Europa's surface isn't as rugged as that of the rockier moons in the solar system. The constant Jovian tidal forces working on the ice and slush tend to smooth out the most extreme differences in height.

But at the micro level, down where a wheeled or tracked element must traverse, the surface resembles an unweathered lava field, filled with sharp boulders, crossed with narrow but deep fissures, cracks, and cycloids. These, of course, were mapped by Element Earl on his original recon—collecting that data was one of his primary goals, so it could be beamed to earth, turned into a three-dimensional map file, then uplinked to Element Rebecca.

The problem is, new cycloids can form in days, changing the whole landscape. Before Element Rebecca, her traverse delayed due to other equipment problems, gets five kilometers from Hoppa, her map ceases to be useful.

And there she stops, asking for guidance.

Earl Tolan is what they used to call an unsympathetic character, back when people still made such judgments. You wouldn't like him, on first meeting. He is smart and also opinionated, a combination which has made friends, family, and co-workers uncomfortable, since he has a bad habit of telling others how best to live their lives, and with great accuracy.

You could wonder—Earl does, in his rare reflective moments—whether this trait was magnified by his twenty years in space ops, where you don't open your mouth unless you're sure of your facts, or Earl prospered in that field because it suited his nature.

He's also bullheaded and fatalistic. See above.

He has paid for his sins, however, in two failed marriages and the cool, distant relationships with his three children. His first marriage, to Kerry, the girl from his hometown in Tennessee, crumbled under the weight of too many moves, too much travel, ridiculous working hours. Kerry, who had put her own career on hold, understandably resented raising three children by herself. Earl, even less sympathetic in this period of his life than at present, started a relationship with Jilliane, a co-worker, which destroyed the marriage as quickly and thoroughly as if targeted by a cruise missile.

The collateral damage was to Earl's relationship with his three children, aged twelve, ten, and seven at the time of the breakup. His oldest daughter, Jordan, decided that the divorce was probably only seventy-five percent Earl's fault, and managed to forgive him, and even made friends with Jilliane when she and Earl married.

But the younger two children, Ben and Marcy, were lost

to Earl. They are cordial, exchanging Christmas cards and the occasional phone call, and possibly seeing each other every two years. But their lives no longer intersect.

Jordan, who is in touch with her father more frequently, saw what you would see, if you spent time with Earl. His energy, for example. It is formidable enough when employed on a project such as J^2E^2, but is downright memorable when put to use on, say, a weekend vacation with Jordan and her family, or on a remodeling job at her small house in Tucson.

Maybe this will help: Earl has learned some of life's harsher lessons. He works less. He flosses more often. He no longer allows a first impression to be his only impression.

"Guess what. We have a problem."

It is the day after the cute meet in the AGC parking lot. On the floor below J^2E^2 mission control, Earl is buttoning his shirt after a shower and pro forma medical check, having just pulled the maximum authorized SLIPPER shift in taking Element Earl back to Hoppa Station. Gareth Haas, the Swiss deputy flight director, shows up. With him is Rebecca Marceau, half out of her SLIPPER suit. She is sweaty, her skin is lined with smeared marks from suit sensors, and her green eyes are red. At first Earl is almost disgusted by the sight of her.

Then he tries to be charitable, knowing that he wasn't looking any better half an hour earlier, knowing that, let's face it, in physical terms, with his stocky build, thinning hair, thick jaw, and heavy brows, he's not much of a prize on his best day.

Especially with the results of his tests, just received this morning before his shift.

"I'm listening."

Haas and Rebecca explain the difficulties. "Rebecca," he says, meaning Element Rebecca, "can't get to the site."

Earl feels sick to his stomach. "Something wrong with the map?" The map derived from Element Earl data.

"The map's perfect," Rebecca says. "But Tufts Passage seems to have gotten tighter." She is referring to a tunnel in an ice hill just large enough for Element Earl (which is, in fact, about the size of a supermarket shopping cart) to pass through. "I'm stuck. Can't go forward, can't back up."

"That's pretty goddamn strange," Earl says.

"It might have been something as simple as the heat of Earl's passage melting the ice," Haas says, trying to be helpful.

The power module's right on my butt, too," Rebecca says, "and Asif's even fatter than I am." She means Element Asif, named for its operator, a Bangladeshi Earl doesn't know well.

"So you need me to map a new route." What Earl wants to do is walk out of J^2E^2 mission control and never look back. To go to his forty-five-footer and take a sail, and maybe never come back. But what he says is, "Let's do it."

"You're outside your margin," Haas says. "I can't ask you to do the job."

"I'll get the doctors to sign a waiver."

"They won't. You know that."

"It's so risky," Rebecca says. "What if he has a failure while you're linked." This was a genuine problem: Ten years ago, during an earlier AGC SLIPPER operation on Mars, an operator happened to be linked real-time when his rover suffered a catastrophic failure. The operator suffered a stroke and was never the same again. Hence the limits and mission rules.

"Earl won't let me down," Earl says.

"He's got all the power he needs," Haas says, agreeing, "but he's had the Big Chill. He'll be going back into the cold without a bake. The accident rate is substantially higher—"

"I know that, you know that, we all know that," Earl snaps. "We also know that you wouldn't have asked me if you didn't need me. So let's go."

Rebecca requires further convincing. "What about the doctors?"

"Don't tell them I'm getting back in the suit."

Angry at their clumsiness, he chases them out of the dressing room. As he begins to don the suit, however, his mood changes. What if something did happen to Element Earl? The human operator knows that a mission is finite, that his linkage won't go on forever. But the elements on Europa are powered by radio-thermal generators that can give life for hundreds of years. Unless an element is totally destroyed, it lives on, diminished, possibly blind, but capable of responding to stimuli or processing data.

He zips up the suit, feeling a surprising pang of sadness. For Element Earl, or himself?

It is always a mixture of pleasure and terror, being linked via SLIPPER to an element on Europa. One of Earl's first instructors, knowing Earl's fondness for sailing and things nautical, compared it to Acapulco cliff diving. After a dozen sessions in the SLIPPER suit, Earl decided that his instructor was an idiot. Linking with an element was only like diving off a cliff if the moment of fear and exhilaration were stretched to an hour. Yes, there is the wonder of feeling that you are crunching Europan snow beneath your "feet," navigating your way through the jumbled heaps of ice like a child picking his way through a forest.

But you must also endure the sheer discomfort of the SLIPPER suit: the data leads that bite and scratch; the sweat that oozes from your neck, armpits, and crotch (occasionally shorting out a lead), then cools to a clammy

pool in the small of your back; the stomach-turning smell
of burnt flesh (which no one can seem to explain); the data
overlays that mar your pristine vision; the goddamn chatter from Haas and his team, who treat all operators like
children with "special needs"—all while feeling that you
are being flung across the universe on the nose of a star-
ship driven at near-light speed by a drunk.

Somehow, Earl forces himself to accept the usual
stresses while ignoring the protests from the medical support team as he drives Element Earl back out on the trail.
(The doctors have been conditioned to look for conditions
that could be linked directly to SLIPPER side effects.
Other than that, they give the operators great license, especially since each operator has already released AGC
from liability now and forever.) For amusement, he
watches the thermal readout of his element's temperature.
It dropped sharply as he exited the Hoppa shelter, and now
it climbs slowly as friction and the general expenditure of
heat are displayed. It reminds Earl of waiting for a download on his first computer forty years back.

Except for the thin wall between booths, Earl and Rebecca
could reach out and touch fingertips. Yet each exchange of
data must go from Earl to Hoppa Station to Element Earl
to Element Rebecca back to Hoppa and La Jolla, a round
trip of 964,000,000 miles in a fraction of a second, thanks
to the SLIPPER technology, which pumps data at 300
times the speed of light. For years Earl grew excited every
time he thought about the process; now, of course, he finds
even the tiniest glitch or lag to be an annoyance.

Today he even finds the traverse of Europa to be less
than totally engaging. He is re-covering the same ground
as the earlier traverse, in essence, crawling through an icy
ditch for the second time.

But then he emerges onto a spot of flat ground, notes
the tracks of Element Rebecca and its power unit on his
original route, and veers off.

This is more challenging, up and down the slopes at an amazing five kilometers an hour. It feels like sailing in the open sea.

Then, just as Earl has grown comfortable with the traverse, Element Earl stalls on a slope that is slightly too steep. He is also in a shadow. Several data packets are squirted back, forth, and around, their tone as close to panic as the operators and mission control ever get. Earl is encouraged to let Element Earl slip backward down the icy slope in search of traction. Meanwhile, the Hoppa base unit will try to find a passable route—

Now the temperature readout, having gotten no higher than a sixth of the way up its scale, starts to plummet, like a barometer just before a storm. Earl finds this troubling, but knows that turning around now would mean doom.

"Back up twenty-two meters," Haas says on the voice loop. "We've got something here."

Element Earl slowly retraces his path—blindly, since the camera only points forward—but surely, since each turn of his wheels has been recorded and can be replayed precisely in reverse. Out of the shadow into the light.

Then forward into what appears to be a narrow passage in a wall of ice. Left. Left again. Temperature rising again. Good. Had it dropped much more, Earl would have had to begin the lengthy disengagement process—

Ping! It's Element Rebecca pulsing him, in direct line of sight. One more turn to the left, and Element Earl has visual, not only on Rebecca, but on Element Asif, the power rover, behind.

There is time for one slight push, an expensive one in terms of power. An electrical arc leaps between them, a common enough event when two machines touch in a vacuum. The event startles both Earls, and causes the displays to drop out for a moment.

Then all is well. Element Rebecca slews free, and continues backing up, clearing the way for Earl to approach Asif. "The drill site is that way. Follow me."

"How do you like the work so far?" Earl has checked into Rebecca's background and knows that the J^2E^2 mission is her first. Just as he knows that her personal history makes him look like a model of stability, with three marriages (none lasting longer than four years) and at least one other semi-famous liaison. No children. Remembering a phrase from his youth, Earl has decided that Rebecca has commitment issues.

"Europa? It reminds me of home."

"You must have grown up someplace very cold and a long time ago." Which is a joke, since by 2026, after thirty years of global warming, there aren't many cold places left on the planet.

"It's not so much the cold," she says. "It's big Jupiter. My parents were teachers in B.C., British Columbia. We lived in a place called Garibaldi, which had this gigantic rock face hanging over it. It always creeped me out. Jupiter feels like that."

They are having martinis as they watch the sun set from the stern of Earl's boat, the *Atropos,* in its slip in Mission Bay. Both have been drained by the experience on Europa today, which required them to operate for six hours in Rebecca's case, ten in Earl's—much longer than the usual three. In spite of his initial feeling that he and Rebecca will never have anything beyond a professional relationship, Earl has accepted her invitation for a drink. A tribute to his stamina, she says.

Hoping to control the agenda, he suggested they come to his boat. Where he pours a second round, as a tribute to her courage, he says, and now Earl is feeling the effects of the alcohol, something he does not enjoy. But he would rather stay here overlooking the Pacific than return to his condo.

"How about you?" she says. "You've been doing this work almost from the beginning."

Earl is not one for introspection or emotion, or so he believes. "It's a great way to be on the cutting edge of exploration at an age when everyone else is retired."

She nods, amused at the banality of this. "Yeah, let's strike a blow for our demo. Age shall not only not wither us, it shan't even slow us down." Then she looks at him closely. "Earl, forgive me, we hardly know each other, but you don't look well."

And then, his barriers eroded by vodka, he starts to weep. "I've got a growth in my neck." In spite of his reservations, he reaches for her, and she takes him in.

During the next week, the elements on Europa move into position. Element Earl stays in Pathfinder mode, blazing a trail to the crevasse picked out years ago by prior orbiting imagers. Element Rebecca follows, and deploys her drilling rig. Element Asif sets up nearby, a portable power station for the submersible operation. And the cargo element begins its trek from Hoppa carrying the submersible that will soon be sinking through Europa's ice into the mysterious darkness below.

The operations run relatively smoothly, with only nagging glitches caused by momentary loss of signal and a few jounces from J-quakes.

Here's the funny thing about elements like Earl and Rebecca: They are only being operated during critical maneuvers, perhaps a few hours out of every twenty-four. The rest of the time, when not powered down or recharging, they are autonomous.

There is a persistent feeling among all operators that their elements retain some of their personalities, even when the link is gone. It's silly, of course. As Earl's idiot instructor once said, "A turned-off lightbulb doesn't remember that it used to give light!" To which Earl, in spite of his agreement with the instructor's point, answered, "A mobile computer with several gigabytes of memory is not a goddamn lightbulb."

Every time Earl and Rebecca go back into operation, they find that Earl, no matter what his last programmed position, has returned to the crevasse where Element Rebecca chews through the ice. "I think it might be a case of love at first bite," Rebecca tells Earl one night, as they walk along the dock, hand in hand.

Earl's response is to kiss her, though he stops a bit sooner than she would like. "I won't break," she tells him, playfully.

"*I* might, though." Earl feels frail, or dishonest. He has told Rebecca everything the doctors told him, that the growth is malignant, but that chemo and radiation and even some experimental genetic treatments might knock it down. For the first few days after being slammed with the news, he almost laughed it off, knowing he could fight and win. But the first rounds of chemo left him shaken. The horizon of his life has drawn closer, like that of an ice plain on Europa compared to the Pacific.

"I'll be gentle," she says, kissing him again. Rebecca's intensity has helped. It's as if she is offering her own strength as another form of treatment.

This is an evening in winter, with the marine layer already rolling in from the west, shrouding the hills of Point Loma across the bay. Earl is lost in them. "Still plowing snow on Europa?" she says, fishing for a connection.

"No. Thinking about a trip I've wanted to make." He nods out to sea. "Catalina Island's out there, a hundred miles away. I've always wanted to sail up and never have."

"Doesn't AGC give vacations?"

"Sure. But nobody wants to take one with an op in progress."

"This one will end."

"For you," he says, meaning Element Rebecca, who only has so much drilling to accomplish before she is shunted off to the side, to a secondary mapping mission for which she is ill-equipped. "Sorry," he adds, realizing how shitty and snappish he sounds. "I just—"

She touches a finger to his lips. "Sssh. I know exactly what you mean. I knew the ops plan when I signed up."

Within a few steps they reach the *Atropos,* and the sight of it bobbing in the twilight raises Earl's spirits. By the time he has finished rigging it for an evening sail, he feels strong enough to face anything, and slightly ashamed of his earlier weakness. "Love at first *byte*," he says, laughing. "I just now got it."

As the drilling proceeds, Element Earl is relegated to geological surveys of the area further to the north and east of the site. He finds it smoother, icier, and flatter than the terrain around Hoppa Station, and Earl himself wonders again why that location was chosen, only to be told by Haas that it provided easier access to the crevasse. Or so it seemed.

In any case, the flight control team and the science support group are completely consumed by the descent of the submersible element through the ice and "the beginnings of the first real search for life in the history of human exploration of the solar system"—at least, according to the AGC Website.

The cargo unit has replaced Element Rebecca at the drillhead, and she has been moved off to her secondary mission as well, mapping to the south and east of the hole in the ice, her data combined with Element Earl's to give a multidimensional picture of the terrain. They amuse themselves by giving completely inappropriate southern California names to Europan landmarks: Point Loma for an ice lake, the Beach and Tennis Club for a jumble of ice boulders, Angeles Crest for a jagged crevasse, Catalina Island for a passageway visible on the far end of Point Loma.

Neither element can venture too far away, of course, since they need to be in line-of-sight comm every few hours. Whenever Earl suits up, he finds himself strangely comforted by the sight of Element Rebecca—shiny, boxlike, asymmetrical, and small—through Element Earl's sensors.

In between shifts, Earl deals with ex-wives Kerry and Jilliane. The old bitterness toward and from Kerry still garbles communications between them, the way a solar flare degrades the SLIPPER link. The fact of Earl's new condition only means that Kerry will allow some sympathy and tenderness to leak into encounters that have been frosty for years. The same applies to the children Ben and Marcy.

Jilliane, who ultimately left Earl four years ago, is consumed by guilt, and offers herself as everything from nurse to sexual partner, until Earl's work schedule and general moodiness cause her to remember why she ran off in the first place. Rebecca's presence makes her feel superfluous.

Then there is Jordan, who takes time from her family and flies to La Jolla for a visit. She meets Rebecca, and offers her approval, and will be present whenever Earl needs her. At the moment, that's not often. He believes he will beat the disease—at least postponing his inevitable doom by five years.

A month to the day after meeting Rebecca, after his diagnosis, Earl shows up at AGC mission control with his head shaved. Concerned about his privacy, and surprised, Rebecca can't ask him why until hours later.

"I start chemo on Monday," Earl says, tentatively rubbing his shiny dome. "The hair is going to be the first casualty."

"Not right away!" she says, protesting.

"No. But everyone will be able to see it coming out in clumps, and I'd rather not display my deterioration so soon."

Rebecca's despair over Earl's change in looks—the pale, naked skull is not an improvement—and Earl's own ambivalence over what may have been a self-destructive impulse are lost in the broad spectrum noise emerging from the science support room at AGC mission control. The submersible element, after three weeks of increasingly frustrating dives in the lightless freezing slurry that is Europa-under-the-crust, has picked up motion at the very limited of its sonar system.

Is it some sort of animal or plant life? Or is it a spurious signal? The science team and its journalistic symbionts spread the news anyway.

When Earl and Rebecca return to AGC early the next day for their shifts, they are forced to park off the site and walk through the crowd that has gathered.

Earl, just out of a chemo session, is weakened by the walk and the wait to a degree he finds astonishing. He barely has the strength to zip up his SLIPPER suit, alarming the medical support team, who know by now that he has a "problem."

Even Rebecca finds herself distracted and jittery when she finally dons her SLIPPER suit to resume the mapping operation.

It is Element Rebecca and Element Earl who find themselves together on the Europan ice plain. "Just imagine," Rebecca says, thumping one of her manipulators on the surface, "something is swimming around down there."

"Yeah, the submersible."

"Come on! I mean some Europan jellyfish! Doesn't that excite you?"

"Only because it means we accomplished the mission."

"That's not very romantic."

"Who said I was romantic?"

"*You* did. You and your blue eyes and your goddamned boat and sailing to Catalina—"

"Well, I'm not feeling very romantic these days. Unless dying of the same disease that killed U.S. Grant and Babe Ruth is romantic."

In La Jolla, Rebecca forms an answer, but even at three hundred-plus times light speed, there is not enough time to relay it, because Element Rebecca has rolled across a thin sheet of ice insufficient to support even a mass of twenty kilograms.

The ice cracks, separates. As Element Earl helplessly records the scene from a distance of sixty-five meters, Element Rebecca teeters on the fissure, antenna slewing one way, the drilling arm swinging forward in what can only be a desperate search for traction, then silently disappears into a crevasse.

The aftermath of the event is prolonged and messy. There is only momentary loss of comm between Rebecca and her element, because Element Earl moves into position at the rim of the crevasse and provides line-of-sight.

Rebecca herself experiences the loss of support and the beginning of a terrifying plunge just as surely as if she'd been standing on the Europan ice in person.

Then there is nothing.

Then there comes a rattle of almost randomly-scattered data bits, quickly telling Rebecca that her element is wedged on its side in a fissure of ice, that her drilling arm and camera have been torn off. She is blind, broken, beyond reach.

But alive. Her radio-thermal power source ensures that

Element Rebecca will continue to send data for the next several years.

Nauseous from his medication and the horrifying accident, Earl can do nothing but wait, though not silently. Even while operating Element Earl, he has grown irritated with the mission control team's obvious distraction, as the ghost sonar squiggle of a theoretical Europan life form is played over and over again. "Haas," he snaps on the open loop, "drop the Ahab routine and pay some fucking attention here."

"No need to get nasty, Earl," Haas says. "We're on top of things."

"If you were on top of things, she wouldn't have fallen."

"Earl," Rebecca says. "It's okay."

Hearing her voice quiets him, as does the false serenity of the Europan landscape. Jupiter is at the edge of his field of vision. The sight angers him. Big fat useless ball of ice—

Then he sees nothing at all. The link between Element Earl and La Jolla still functions, but the La Jolla end has failed.

Earl Tolan is taken to UC-SD Medical Center, where he dies four hours later. The cause of death is listed as a heart attack; the real cause is almost certainly complications from throat cancer and related treatment.

Once over her shock at the double loss of a single day— Element Rebecca and Earl himself—Rebecca sees the unexpected heart attack as a blessing, saving Earl and Rebecca and Jordan the horror of the almost certain laryngectomy and talking through a stoma and more radiation and the swelling and the pain and the horror of knowing that it will never get better, only worse.

Rebecca helps Jordan dispose of Earl's possessions.

The *Atropos* is the trickiest of them, ultimately sold for a pittance in a depressed boating market.

The submersible element records more ghost blips before falling silent, a victim of cold, several weeks past its design life. Rebecca resigns from the operator program and is reassigned to AGC's "advanced planning" unit, helping with the design of a new set of elements for another Europan mission.

One day three months after that awful day she returns to mission control, dons a SLIPPER suit, and spends a few moments on the icy plains of Europa with Element Earl.

Her last command aims him across Point Loma toward distant Catalina.

NEVERMORE

Ian R. MacLeod

British writer Ian R. MacLeod was one of the hottest new writers of the '90s, and, as the new century begins, his work continues to grow in power and deepen in maturity. MacLeod has published a slew of strong stories in Interzone, Asimov's Science Fiction, Weird Tales, Amazing, *and* The Magazine of Fantasy & Science Fiction, *among other markets. Several of these stories made the cut for one or another of the various "Best of the Year" anthologies; in 1990, in fact, he appeared in* three *different Best of the Year anthologies with three different stories, certainly a rare distinction. His first novel,* The Great Wheel, *was published to critical acclaim in 1997, followed by a major collection of his short work,* Voyages by Starlight. *In 1999, he won the World Fantasy Award with his brilliant novella "The Summer Isles," and followed it up in 2000 by winning another World Fantasy Award for his novelette "The Chop Girl." MacLeod lives with his wife and young daughter in the West Midlands of England, and is at work on several new novels.*

In the stylish and compelling story that follows, he shows us that Passion can persist beyond flesh . . . but that you always have to worry about drifting apart in a relationship—even when one of you is dead.

Now that he couldn't afford to buy enough reality, Gustav had no option but to paint what he saw in his dreams. With no sketchpad to bring back, no palette or cursor, his head rolling up from the pillow and his mouth dry and his

jaw aching from the booze he'd drunk the evening before—which was the cheapest means he'd yet found of
getting to sleep—he was left with just that one chance, and
a few trailing wisps of something that might once have
been beautiful before he had to face the void of the day.

It hadn't started like this, but he could see by now that
this was how it had probably ended. Representational art
had had its heyday, and for a while he'd been feted like the
bright new talent he'd once been sure he was. And big
lumpy actuality that you could smell and taste and get
under your fingernails would probably come back into
style again—long after it had ceased to matter to him.

So that was it. Load upon load of self-pity falling down
upon him this morning from the damp-stained ceiling.
What *had* he been dreaming? Something—surely something. Otherwise being here and being Gustav wouldn't
come as this big a jolt. He should've gotten more used to
it than this by now. . . . Gustav scratched himself, and discovered that he also had an erection, which was another
sign—hadn't he read once, somewhere?—that you'd been
dreaming dreams of the old-fashioned kind, unsimulated,
unaided. A sign, anyway, of a kind of biological optimism.
The hope that there might just be a hope.

Arthritic, Cro-Magnon, he wandered out from his bed.
Knobbled legs, knobbled veins, knobble toes. He still
missed the habit of fiddling with the controls of his window in the pockmarked far wall, changing the perspectives
and the light in the dim hope that he might stumble across
something better. The sun and the moon were blazing
down over Paris from their respective quadrants, pouring
like mercury through the nanosmog. He pressed his hand
to the glass, feeling the watery wheeze of the crack that
now snaked across it. Five stories up in these scrawny
empty tenements, and a long, long way down. He laid his
forehead against its coolness as the sour thought that he
might try to paint this scene speeded through him. He'd

finished at least twenty paintings of foreal Paris; all reality engines and cabled ruins in gray, black, and white. Probably done, oh, at least several hundred studies in inkwash, pencil, charcoal. No one would ever buy them, and for once they were right. The things were passionless, ugly—he pitied the potentially lovely canvases he'd ruined to make them. He pulled back from the window and looked down at himself. His erection had faded from sight beneath his belly.

Gustav shuffled through food wrappers and scrunched-up bits of cartridge paper. Leaning drifts of canvas frames turned their backs from him toward the walls, whispering on breaths of turpentine of things that might once have been. But that was okay, because he didn't have any paint right now. Maybe later, he'd get the daft feeling that, today, something might work out, and he'd sell himself for a few credits in some stupid trick or other—what had it been last time; painting roses red dressed as a playing card?—and the supply ducts would bear him a few precious tubes of oils. And a few hours after that he'd be—But what was that noise?

A thin white droning like a plastic insect. In fact, it had been there all along—had probably woken him at this ridiculous hour—but had seemed so much a part of everything else that he hadn't noticed. Gustav looked around, tilting his head until his better ear located the source. He slid a sticky avalanche of canvas board and cotton paper off an old chair, and burrowed in the cushions until his hand closed on a telephone. He'd only kept the thing because it was so cheap that the phone company hadn't bothered to disconnect the line when he'd stopped paying. That was, if the telephone company still existed. The telephone was chipped from the time he'd thrown it across the room after his last conversation with his agent. But he touched the activate pad anyway, not expecting anything more than a blip in the system, white machine noise.

"Gustav, you're still *there,* are you?"

He stared at the mouthpiece. It was his dead ex-wife Elanore's voice.

"What do you want?"

"Don't be like that, Gus. Well, *I* won't be anyway. Time's passed, you know, things have changed."

"Sure, and you're going to tell me next that you—"

"Yes, would like to meet up. We're arranging this party. I ran into Marcel in Venice—he's currently Doge there, you know—and we got talking about old times and all the old gang. And so we decided we were due for a reunion. You've been one of the hardest ones to find, Gus. And then I remembered that old tenement . . ."

"Like you say, I'm still here."

"Still painting?"

"Of *course* I'm still painting! It's what I do."

"That's great. Well—sorry to give you so little time, but the whole thing's fixed for this evening. You won't *believe* what everyone's up to now! But then, I suppose you've seen Francine across the sky."

"Look, I'm not sure that I—"

"And we're going for Paris, 1890. Should be right up your street. I've splashed out on all-senses. And the food and the drink'll be foreal. So you'll come, won't you? The past is the past, and I've honestly forgotten about much of it since I passed on. Put it into context, anyway. I really don't bear a grudge. So you *will* come? Remember how it was, Gus? Just smile for me the way you used to. And remember . . ."

Of course he remembered. But he still didn't know what the hell to expect that evening as he waited—too early, despite the fact that he'd done his best to be pointedly late—in the virtual glow of a pavement café off the Rue St-Jacques beneath a sky fuzzy with Van Gogh stars.

Searching the daubed figures strolling along the cobbles, Gustav spotted Elanore coming along before she saw him. He raised a hand, and she came over, sitting down on a wobbly chair at the uneven whirl of the table. Doing his best to maintain a grumpy pose, Gustav called the waiter for wine, and raised his glass to her with trembling fingers. He swallowed it all down. Just as she'd promised, the stuff was foreal.

Elanore smiled at him. And Elanore looked beautiful. Elanore was dressed for the era in a long dress of pure ultramarine. Her red hair was bunched up beneath a narrow-brimmed hat adorned with flowers.

"It's about now," she said, "that you tell me I haven't changed."

"And you tell me that I *have*."

She nodded. "But it's true. Although you haven't changed *that* much, Gus. You've aged, but you're still one of the most . . . solid people I know."

Elanore offered him a Disc Bleu. He took it, although he hadn't smoked in years and she'd always complained that the things were bad for him when she was alive. Elanore's skin felt cool and dry in the moment that their hands touched, and the taste of the smoke as it shimmered amid the brush strokes was just as it had always been. Music drifted out from the blaze of the bar where dark figures writhed as if in flames. Any moment now, he knew, she'd try to say something vaguely conciliatory, and she'd interrupt as he attempted to do the same.

He gestured around at the daubs and smears of the other empty tables. He said, "I thought I was going to be late. . . ." The underside of the canopy that stretched across the pavement blazed. How poor old Vincent had loved his cadmiums and chromes! And never sold one single fucking painting in his entire life.

"What—what I told you was true," Elanore said, stumbling slightly over these little words, sounding almost un-Elanore-like for a moment; nearly uneasy. "I mean, about Marcel in Venice and Francine across the sky. And, yes, we *did* talk about a reunion. But you know how these things are. Time's precious, and, at the end of the day it's been so long that these things really do take a lot of nerve. So it didn't come off. It was just a few promises that no one really imagined they'd keep. But I thought—well, I thought that it would be nice to see *you* anyway. At least one more time."

"So all of this is just for me. *Jesus,* Elanore, I knew you were rich, but . . ."

"Don't be like that, Gustav. I'm not trying to impress you or depress you or whatever. It was just the way it came out."

He poured more of the wine, wondering as he did so exactly what trick it was that allowed them to share it.

"So, you're still painting?"

"Yep."

"I haven't seen much of your work about."

"I do it for private clients," Gustav said. "Mostly."

He glared at Elanore, daring her to challenge his statement. Of course, if he really *was* painting and selling, he'd have some credit. And if he had *credit,* he wouldn't be living in that dreadful tenement she'd tracked him down to. He'd have paid for all the necessary treatments to stop himself becoming the frail old man he so nearly was. *I can help, you know,* Gustav could hear Elanore saying, because he'd heard her say it so many times before. *I don't need all this wealth. So let me give you just a little help. Give me that chance. . . .* But what she actually *said* was even worse.

"Are you recording yourself, Gus?" Elanore asked. "Do you have a librarian?"

Now, he thought, now is the time to walk out. Pull this whole thing down and go back into the street—the foreal street. And forget.

"Did you know," he said instead, "that the word *reality* once actually *meant* foreal—not the projections and the simulations, but proper actuality. But then along came *virtual* reality, and of course, when the *next* generation of products was developed, the illusion was so much better that you could walk right into it instead of having to put on goggles and a suit. So they had to think of an improved phrase, a super-word for the purposes of marketing. And someone must have said, *Why don't we just call it reality?*"

"You don't have to be hurtful, Gus. There's no rule written down that says we can't get on."

"I thought that that was exactly the problem. It's in my head, and it was probably there in yours before you died. Now it's . . ." He'd have said more. But he was suddenly, stupidly, near to tears.

"What exactly *are* you doing these days, Gus?" she asked as he cleared his throat and pretended it was the wine that he'd choked on. "What are you painting at the moment?"

"I'm working on a series," he was surprised to hear himself saying. "It's a sort of a journey-piece. A sequence of paintings which began here in Paris and then . . ." He swallowed. ". . . bright, dark colors . . ." A nerve began to leap beside his eye. Something seemed to touch him, but was too faint to be heard or felt or seen.

"Sounds good, Gus," Elanore said, leaning toward him across the table. And Elanore smelled of Elanore, the way she always did. Her pale skin was freckled from the sunlight of whatever warm and virtual place she was living. Across her cheeks and her upper lip, threaded gold, lay the down that he'd brushed so many times with the tips of his fingers. "I can tell from that look in your eyes that you're into a real good phase. . . ."

After that, things went better. They shared a second bottle of *vin ordinaire*. They made a little mountain of the butts of her Disc Bleu in the ashtray. This ghost—she really *was* like Elanore. Gustav didn't even object to her taking his hand across the table. There was a kind of abandon in all of this—new ideas mixed with old memories. And he understood more clearly now what Van Gogh had meant about this café being a place where you could ruin yourself, or go mad or commit a crime.

The few other diners faded. The virtual waiters, their aprons a single assured gray-white stroke of the palette knife, started to tip the chairs against the tables. The aromas of the Left Bank's ever-unreliable sewers began to override those of cigarettes and people and horse dung and wine. At least, Gustav thought, *that* was still foreal. . . .

"I suppose quite a lot of the others have died by now," Gustav said. "All that facile gang you seem to so fondly remember."

"People still change, you know. Just because we've passed on, doesn't mean we can't *change.*"

By now, he was in a mellow enough mood just to nod at that. And how have *you* changed, Elanore? he wondered. After so long, what flicker of the electrons made you decide to come to me now?

"You're obviously doing well."

"I am . . ." She nodded, as if the idea surprised her. "I mean, I didn't expect—"

"And you look—"

"And *you,* Gus, what I said about you being—"

"That project of mine—"

"I know, I—"

They stopped and gazed at each other. Then they both smiled, and the moment seemed to hold, warm and frozen, as if from a scene within a painting. It was almost . . .

"Well . . ." Elanore broke the illusion first as she began to fumble in the small sequined purse she had on her lap. Eventually, she produced a handkerchief and blew delicately on her nose. Gustav tried not to grind his teeth—although this was *exactly* the kind of affectation he detested about ghosts. He guessed, anyway, from the changed look on her face, that she knew what he was thinking. "I suppose that's it, then, isn't it, Gus? We've met—we've spent the evening together without arguing. Almost like old times."

"Nothing will ever be like old times."

"No . . ." Her eyes glinted, and he thought for a moment that she was going to become angry—goaded at last into something like the Elanore of old. But she just smiled. "Nothing ever will be like old times. That's the problem, isn't it? Nothing ever was, or ever will be . . ."

Elanore clipped her purse shut again. Elanore stood up. Gustav saw her hesitate as she considered bending down to kiss him farewell, then decided that he would just regard that as another affront, another slap in the face.

Elanore turned and walked away from Gustav, fading into the chiaroscuro swirls of lamplight and gray.

Elanore, as if Gustav needed reminding, had been alive when he'd first met her. In fact, he'd never known anyone who was *more* so. Of course, the age difference between them was always huge—she'd already been past a hundred by then, and he was barely forty—but they'd agreed on that first day that they met, and on many days after, that there was a corner in time around which the old eventually turned to rejoin the young.

In another age, and although she always laughingly denied it, Gustav always suspected that Elanore would have had her sagging breasts implanted with silicone, the wrinkles stretched back from her face, her heart replaced by a throbbing steel simulacrum. But she was lucky enough to exist at a time when effective antiaging treatments were finally available. As a post-centarian, wise and rich and moderately, pleasantly, famous, Elanore was probably more fresh and beautiful than she'd been at any other era in her life. Gustav had met her at a party beside a Russian lake—guests wandering amid dunes of snow. Foreal had been a fashionable option then; although for Gustav, the grounds of this pillared ice-crystaled palace that Catherine the Great's Scottish favorite Charles Cameron had built seemed far too gorgeous to be entirely true. But it *was* true—foreal, actual, concrete, genuine, unvirtual—and such knowledge was what had driven him then. That, and the huge impossibility of ever really managing to convey any of it as a painter. That, and the absolute certainty that he would *try*.

Elanore had wandered up to him from the forest dusk dressed in seal furs. The shock of her beauty had been like all the rubbish he'd heard other artists talk about and thus so detested. And he'd been a stammering wreck, but somehow that hadn't mattered. There had been—and here again the words became stupid, meaningless—a dazed physicality between them from that first moment that was so intense, it was spiritual.

Elanore told Gustav that she'd seen and admired the series of triptychs he'd just finished working on. They were painted directly onto slabs of wood, and depicted totemistic figures in dense blocks of color. The critics had generally damned them with faint praise—had talked of Cubism and Mondrian—and were somehow unable to recognize Gustav's obvious and grateful debt to Guaguin's Tahitian paintings. But Elanore had seen and understood those bright muddy colors. And, yes, she'd dabbled a little in painting herself—just enough to know that truly creative acts were probably beyond her . . .

Elanore wore her red hair short in those days. And there were freckles, then as always, scattered across the bridge of her nose. She showed the tips of her teeth when she smiled, and he was conscious of her lips and her tongue. He could smell, faint within the clouds of breath that entwined them, her womanly scent.

A small black cat threaded its way between them as they talked, then, barely breaking the crust of the snow, leapt up onto a bough of the nearest pine and crouched there, watching them with emerald eyes.

"That's Metzengerstein," Elanore said, her own even greener eyes flickering across Gustav's face, but never ceasing to regard him. "He's my librarian."

When they made love later on in the agate pavilion's frozen glow, and as the smoke of their breath and their sweat clouded the winter twilight, all the disparate elements of Gustav's world finally seemed to join. He carved Elanore's breasts with his fingers and tongue, and painted her with her juices, and plunged into her sweet depths, and came, finally, finally, and quite deliciously, as her fingers slid around and he in turn was parted and entered by her.

Swimming back up from that, soaked with Elanore, exhausted, but his cock amazingly still half-stiff and rising, Gustav became conscious of the black cat that all this time had been threading its way between them. Its tail now curled against his thigh, corrugating his scrotum. Its claws gently kneaded his belly.

Elanore had laughed and picked Metzengerstein up, purring herself as she laid the creature between her breasts.

Gustav understood. Then or later, there was never any need for her to say more. After all, even Elanore couldn't live forever—and she needed a librarian with her to record her thoughts and actions if she was ever to pass on. For all its myriad complexities, the human brain had evolved to last a single lifetime; after that, the memories and impressions eventually began to overflow, the data became corrupted. Yes, Gustav understood. He even came to like the way Metzengerstein followed Elanore around like a witch's familiar, and, yes, its soft, sharp cajolings as they made love.

Did they call them ghosts then? Gustav couldn't remember. It was a word, anyway—like spic, or nigger—

that you never used in front of them. When he and Elanore
were married, when Gustav loved and painted and loved
and painted her, when she gave him her life and her spirit
and his own career somehow began to take off as he finally
mastered the trick of getting some of the passion he felt
onto the lovely, awkward canvas, he always knew that part
of the intensity between them came from the age gap, the
difference, the inescapable fact that Elanore would soon
have to die.

It finally happened, he remembered, when he was leaving Gauguin's tropic dreams and nightmares behind and
toying with a more straightforwardly Impressionist phase.
Elanore was modeling for him nude as Manet's *Olympia*.
As a concession to practicalities and to the urgency that
then always possessed him when he was painting, the
black maidservant bearing the flowers in his lavish new
studio on the Boulevard des Capucines was a projection,
but the divan and all the hangings, the flowers, and the cat,
of course—although by its programmed nature, Metzengerstein was incapable of looking quite as scared and
scrawny as Manet's original—were all foreal.

"You know," Elanore said, not breaking pose, one hand
toying with the hem of the shawl on which she was lying,
the other laid negligently, possessively, without modesty,
across her pubic triangle, "we really should reinvite Marcel over after all he's done for us lately."

"Marcel?" In honesty, Gustav was paying little attention to anything at that moment other than which shade to
swirl into the boudoir darkness. He dabbed again onto his
testing scrap. "Marcel's in San Francisco. We haven't seen
him in months."

"Of course . . . silly me."

He finally glanced up again, what could have been moments or minutes later, suddenly aware that a cold silence
had set in. Elanore, being Elanore, never forgot anything.

Elanore was light and life. Now, all her *Olympia*-like poise was gone.

This wasn't like the decay and loss of function that affected the elderly in the days before recombinant drugs. Just like her heart and her limbs, Elanore's physical brain still functioned perfectly. But the effect was the same. Confusions and mistakes happened frequently after that, as if consciousness drained rapidly once the initial rent was made. For Elanore, with her exquisite dignity, her continued beauty, her companies and her investments and the contacts that she needed to maintain, the process of senility was particularly terrible. No one, least of all Gustav, argued against her decision to pass on.

Back where reality ended, it was past midnight and the moon was blazing down over the Left Bank's broken rooftops through the grayish brown nanosmog. And exactly where, Gustav wondered, glaring up at it through the still-humming gantries of the reality engine that had enclosed him and Elanore, is Francine across the sky? How much do you have to pay to get the right decoders in your optic nerves to see the stars entwined in some vast projection of her? How much of your life do you have to give away?

The mazy streets behind St. Michael were rotten and weed-grown in the bilious fog, the dulled moonlight. No one but Gustav seemed to live in the half-supported ruins of the Left Bank nowadays. It was just a place for posing in and being seen—although in that respect, Gustav reflected, things really hadn't changed. To get back to his tenement, he had to cross the Boulevard St-Germain through a stream of buzzing robot cars that, no matter how he dodged them, still managed to avoid him. In the busier streets beyond, the big reality engines were still glowing. In fact, it was said that you could now go from one side of

Paris to the other without having to step out into foreal.
Gustav, as ever, did his best to do the opposite, although he
knew that, even without any credit, he would still be freely
admitted to the many realities on offer in these generous,
carefree days. He scowled at the shining planes of the
powerfields that stretched between the gantries like bubbles. Faintly from inside, coming at him from beyond the
humming of the transformers that tamed and organized the
droplets of nanosmog into shapes you could feel, odors
you could smell, chairs you could sit on, he could hear
words and laughter, music, the clink of glasses. He could
even just make out the shapes of the living as they postured
and chatted. It was obvious from the way that they were
grouped that the living were outnumbered by the dead
these days. Outside, in the dim streets, he passed figures
like tumbling decahedrons who bore their own fields with
them as they moved between realities. They were probably
unaware of him as they drifted by, perhaps saw him as
some extra enhancement of whatever dream it was they
were living. Flick, flick. Scheherzade's Baghdad. John
Carter's Mars. It really didn't matter that you were still in
Paris, although Elanore, of course, had showed sensitivity
in the place she had selected for their meeting.

Beyond the last of the reality engines, Gustav's own
cheap unvirtual tenement loomed into view. He picked his
way across the tarmac toward the faint neon of the foreal
Spar store beside it. Inside, there were the usual gray slabs
of packaging with tiny windows promising every possible
delight. he wandered up the aisles and activated the
homely presence of the woman who served the dozen or so
anachronistic places that were still scattered around Paris.
She smiled at him—a living ghost, really; but then, people
seemed to prefer the illusion of the personal touch. Behind
her, he noticed, was an antiquated cigarette machine. He
ordered a packet of Disc Bleu, and palmed what were
probably the last of his credits—which amounted to half a

stick of charcoal or two squeezes-worth of Red Lake. It
was a surprise to him, in fact, that he even had enough for
these cigarettes.

Outside, ignoring the health warning that flashed
briefly before his eyes, he lighted a Disc Bleu, put it to his
lips, and deeply inhaled. A few moments later, he was in a
nauseous sweat, doubled up and gasping.

Another bleak morning, timeless and gray. This ceiling,
these walls. And Elanore . . . Elanore was dead. Gone.

Gustav belched on the wine he was sure that he'd
drunk, and smelled the sickness and the smoke of that foreal Disc Bleu still clinging to him. But there was no trace
of Elanore. Not a copper strand of hair on his shoulder or
curled around his cock, not her scent riming his hands.

He closed his eyes and tried to picture a woman in a
white chemise bathing in a river's shallows, two bearded
men talking animatedly in a grassy space beneath the trees,
and Elanore sitting naked close by, although she watches
rather than joins in their conversation. . . .

No. That wasn't it.

Somehow getting up, pissing cloudily into the appropriate receptacle, Gustav finally grunted in unsurprise
when he noticed a virtual light flickering through the
heaped and broken frames of his easels. Unlike the telephone, he was sure that the company had disconnected his
terminal long ago. His head fizzing, his groin vaguely
tumescent, some lost bit of the night nagging like a stray
scrap of meat between his teeth, he gazed down into the
spinning options that the screen offered.

It was Elanore's work, of course—or the ghost of en-
tangled electrons that Elanore had become. Hey presto!—
Gustav was back on line; granted this shimmering link into
the lands of the dead and the living. He saw that he even
had positive credit, which explained why he'd been able to

buy that packet of Disc Bleu. He'd have slammed his fist down into the thing if it would have done any good.

Instead, he scowled at his room, the huddle backs of the canvases, the drifts of discarded food and clothing, the heap of his bed, wondering if Elanore was watching him now, thrusting a spare few gigabytes into the sensors of some nano-insect that was hovering close behind him. Indeed, he half-expected the thin partitions and dangling wires, all the mocking rubbish of his life, to shudder and change into snowy Russian parkland, a wooded glade, even Paris again, 1890. But none of that happened.

The positive credit light still glowed enticingly within the terminal. In the almost certain knowledge that he would regret it, but quite unable to stop himself, Gustav scrolled through the pathways that led him to the little-frequented section dealing with artists' foreal requisites. Keeping it simple—down to fresh brushes, and Lefranc and Bourgeois's extra fine Flake White, Cadmium Yellow, Vermilion, Deep Madder, Cobalt Blue, and Emerald Green—and still waiting as the cost all of that clocked up for the familiar credit-expired sign to arrive, he closed the screen.

The materials arrived far more quickly than he'd expected, disgorging themselves into a service alcove in the far corner with a whoosh like the wind. The supplier had even remembered to include the fresh bottles of turpentine he'd forgotten to order—he still had plenty of clean stretched canvases anyway. So here (the feel of the fat new tubes, the beautiful, haunting names of the colors, the faint stirring sounds that the brushes made when he tried to lift them) was everything he might possibly need.

Gustav was an artist.

•   •   •

The hours did funny things when Gustav was painting—or even thinking about painting. They ran fast or slow, passed by on a fairy breeze, or thickened and grew huge as megaliths, then joined up and began to dance lumberingly around him, stamping on every sensibility and hope.

Taking fierce drags of his last Disc Bleu, clouding his tenement's already filmy air, Gustav finally gave up scribbling on his pad and casting sidelong glances at the canvas as the blazing moon began to flood Paris with its own sickly version of evening. As he'd always known he'd probably end up doing, he then began to wander the dim edges of his room, tilting back and examining his old, unsold, and generally unfinished canvases. Especially in this light, and seen from upside down, the scenes of foreal Paris looked suitably wan. There was so little to them, in fact, such a thinness and lack of color, that they could easily be reused. But here in the tangled shadows of the farthest corner, filled with colors that seemed to pour into the air like a perfume, lay his early attempts at Symbolism and Impressionism. . . . Amid those, he noticed something paler again. In fact, unfinished—but from an era when, as far as he could recall, he'd finished everything. He risked lifting the canvas out, and gazed at the outlines, the dabs of paint, the layers of wash. He recognized it now. It had been his attempt at Manet's *Olympia*.

After Elanore had said her good-byes to all her friends, she retreated into the white virtual corridors of a building near the Cimetière du Père Lachaise that might once have been called a hospital. There, as a final fail-safe, her mind was scanned and stored, the lineaments of her body were recorded. Gustav was the only person Elanore allowed to visit her during those last weeks; she was perhaps already too confused to understand what seeing her like this was doing to him. He'd sit amid the webs of silver monitoring

wires as she absently stroked Metzengerstein, and the cat's
eyes, now far greener and brighter than hers, regarded him.
She didn't seem to want to fight this loss of self. That was
probably the thing that hurt him most. Elanore, the proper
foreal Elanore, had always been searching for the next
river to cross, the next challenge; it was probably the one
characteristic that they had shared. But now she accepted
death, this loss of Elanore, with nothing but resignation.
This is the way it is for all of us, Gustav remembered her
saying in one of the last cogent periods before she forgot
his name. *So many of our friends have passed on already.
It's just a matter of joining them. . . .*

Elanore never quite lost her beauty, but she became like
a doll, a model of herself, and her eyes grew vacant as she
sat silent or talked ramblingly. The freckles faded from her
skin. Her mouth grew slack. She began to smell sour.
There was no great fuss made when they finally turned her
off, although Gustav still insisted that he be there. It was a
relief, in fact, when Elanore's eyes finally closed and her
heart stopped beating, when the hand he'd placed in his
turned even more flaccid and cold. Metzengerstein gave
Gustav one final glance before it twisted its way between
the wires, leapt off the bed, and padded from the room, its
tail raised. For a moment, Gustav considered grabbing the
thing, slamming it down into a pulp of memory circuits
and flesh and metal. But it had already been depro-
grammed. Metzengerstein was just a shell; a comforter for
Elanore in her last dim days. He never saw the creature
again.

Just as the living Elanore had promised, her ghost only
returned to Gustav after a decent interval. And she made
no assumptions about their future at that first meeting on
the neutral ground of a shorefront restaurant in virtual Bal-
bec. She clearly understood how difficult all this was for
him. It had been a windy day, he remembered, and the
tablecloths flapped, the napkins threatened to take off, the

lapel of the ream brocade jacket she was wearing kept flying across her throat until she pinned it back with a brooch. She told him that she still loved him, and that she hoped they would be able to stay together. A few days later, in a room in the same hotel overlooking the same windy beach, Elanore and Gustav made love for the first time since she had died.

The illusion, Gustav had to admit, then and later, was always perfect. And, as the dying Elanore had pointed out, they both already knew many ghosts. There was Marcel, for instance, and there was Jean, Gustav's own dealer and agent. It wasn't as if Elanore had even been left with any choice. In a virtual, ghostly daze himself, Gustav agreed that they should set up home together. They chose Brittany, because it was new to them—unloaded with memories— and the scenery was still often decent and visible enough to be worth painting.

Foreal was going out of style by then. For many years, the technologies of what was called reality had been flawless. But now, they became all-embracing. It was at about this time, Gustav supposed, although his memory once again was dim on this matter, that they set fire to the moon. The ever-bigger reality engines required huge amounts of power—and so it was that the robot ships set out, settled into orbit around the moon, and began to spray the surface with antimatter, spreading their wings like hands held out to a fire to absorb and then transmit back to earth the energies this iridescence gave. The power the moon now provided wasn't quite limitless, but it was near enough. With so much alternative joy and light available, the foreal world, much like a garden left untended, soon began to assume a look of neglect.

Ever-considerate to his needs, Elanore chose and had refurbished a gabled clifftop mansion near Locronan, and ordered graceful and foreal furniture at huge extra expense. For a month or so, until the powerlines and trans-

formers of the reality engines had been installed, Gustav
and Elanore could communicate with each other only by
screen. He did his best to tell himself that being unable to
touch her was a kind of tease, and kept his thoughts away
from such questions as where exactly Elanore was when
she wasn't with him, and if she truly imagined she was the
seamless continuation of the living Elanore that she
claimed herself to be.

The house smelled of salt and old stone, and then of wet
plaster and new carpets, and soon began to look as charming and eccentric as anything Elanore had organized in her
life. As for the cost of all this forgotten craftsmanship,
which even in these generous times was quite daunting,
Elanore had discovered, like many of the ghosts who had
gone before her, that her work—the dealing in stocks,
ideas, and raw megawatts in which she specialized—was
suddenly much easier. She could flit across the world,
make deals based on long-term calculations that no living
person could ever hope to understand.

Often, in the early days when Elanore finally reached
the reality of their clifftop house in Brittany, Gustav would
find himself gazing at her, trying to catch her unawares, or,
in the nights when they made love with an obsessive frequency and passion, he would study her whilst she was
sleeping. If she seemed distracted, he put it down to some
deal she was cooking, a new antimatter trial across the Sea
of Storms, perhaps, or a business meeting in Capetown. If
she sighed and smiled in her dreams, he imagined her in
the arms of some long-dead lover.

Of course, Elanore always denied such accusations. She
even gave a good impression of being hurt. She was, she
insisted, configured to ensure that she was always exactly
where she appeared to be, except for brief times and in the
gravest of emergencies. In the brain or on the net, human
consciousness was a fragile thing—permanently in danger
of dissolving. *I really* am *talking to you now, Gustav.* Oth-

erwise, Elanore maintained, she would unravel, she would cease to be Elanore. As if, Gustav thought in generally silent rejoinder, she hadn't ceased to be Elanore already.

She'd changed, for a start. She was cooler, calmer, yet somehow more mercurial. The simple and everyday motions she made, like combing her hair or stirring coffee, began to look stiff and affected. Even her sexual preferences had changed. And passing over *was* different. Yes, she admitted that, even though she could feel the weight and presence of her own body just as she could feel his when he touched her. Once, as the desperation of their arguments increased, she even insisted on stabbing herself with a fork, just so that he might finally understand that she felt pain. But for Gustav, Elanore wasn't like the many other ghosts he'd met and readily accepted. They weren't *Elanore.* He'd never loved and painted *them.*

Gustav soon found that he couldn't paint Elanore now, either. He tried from sketches and from memory; once or twice he got her to pose. But it didn't work. He couldn't quite lose himself enough to forget what she was. They even tried to complete that *Olympia,* although the memory was painful for both of them. She posed for him as Manet's model, who in truth she did look a little like; the same model who'd posed for that odd scene by the river, *Dejéuner sur l'Herbe.* How, of course, the cat as well as the black maid had to be a projection, although they did their best to make everything else the same. But there was something lost and wan about the painting as he tried to develop it. The nakedness of the woman on the canvas no longer gave off strength and knowledge and sexual assurance. She seemed pliant and helpless. Even the colors grew darker; it was like fighting smoke in a dream.

Elanore accepted Gustav's difficulties with what he sometimes found to be chillingly good grace. She was prepared to give him time. He could travel. She could develop new interests, burrow within the net as she'd always promised herself, and live in some entirely different place.

Gustav began to take long walks away from the house, along remote clifftop paths and across empty beaches, where he could be alone. The moon and the sun sometimes cast their silver ladders across the water. Soon, Gustav thought sourly, there'll be nowhere left to escape *to*. Or perhaps we will *all* pass on, and the gantries and the ugly virtual buildings that all look like the old Pompidou Center will cease to be necessary; but for the glimmering of a few electrons, the world will revert to the way it was before people came. We can even extinguish the moon.

He also started to spend more time in the few parts of their rambling house that, largely because much of the stuff they wanted was hand-built and took some time to order, Elanore hadn't yet had fitted out foreal. He interrogated the house's mainframe to discover the codes that would turn the reality engines off and on at will. In a room filled with tapestries, a long oak table, a vase of hydrangeas, pale curtains lifting slightly in the breeze, all it took was the correct gesture, a mere click of his fingers, and it would shudder and vanish, to be replaced by nothing but walls of mildewed plaster, the faint tingling sensation that came from the receding powerfield. There—then gone. Only the foreal view at the window remained the same. And now, click, and it all came *back* again. Even the fucking vase. The fucking flowers.

Elanore sought him out that day. Gustav heard her footsteps on the stairs, and knew that she'd pretend to be puzzled as to why he wasn't working in his studio.

"*There* you are," she said, appearing a little breathless after her climb up the stairs. "I was thinking—"

Finally scratching the itch that he realized had been tickling him for some time, Gustav clicked his fingers. Elanore—and the whole room, the table, the flowers, the tapestries—flickered off.

He waited—several beats, he really didn't know how long. The wind still blew in through the window. The powerfield hummed faintly, waiting for its next command. He clicked his fingers. Elanore and the room took shape again.

"I thought you'd probably override that," he said. "I imagined you'd given yourself a higher priority than the furniture."

"I could if I wished," she said. "I didn't think I'd need to do such a thing."

"No. I mean, you can just go somewhere else, can't you? Some other room in this house. Some other place. Some other continent . . ."

"I keep telling you. It isn't like that."

"I know. Consciousness is fragile."

"And we're really not that different, Gus. I'm made of random droplets held in a force field—but what are *you?* Think about it. You're made of atoms, which are just quantum flickers in the foam of space, particles that aren't even particles at all. . . ."

Gustav stared at her. He was remembering—he couldn't help it—that they'd made love the previous night. Just two different kinds of ghost; entwined, joining—he supposed that that was what she was saying. And what about my *cock,* Elanore, and all the stuff that gets emptied into you when we're fucking? What the hell do you do with *that?*

"Look, Gus, this isn't—"

"And what do you dream at night, Elanore? What is it that you do when you pretend you're sleeping?"

She waved her arms in a furious gesture that Gustav almost recognized from the Elanore of old. "What the hell do you *think* I do, Gus? I *try* to be human. You think it's easy, do you, hanging on like this? You think I enjoy watching *you* flicker in and out?—which is basically what it's like for me every time you step outside these fields? Sometimes I just wish I . . ."

Elanore trailed off there, glaring at him with emerald eyes. Go on, Gustav felt himself urging her. *Say* it, you phantom, shade, wraith, ghost. Say you wish you'd simply died. But instead, she made some internal command of her own, and blanked the room—and vanished.

It was the start of the end of their relationship.

Many guests came to visit their house in the weeks after that, and Elanore and Gustav kept themselves busy in the company of the dead and the living. All the old crowd, all the old jokes. Gustav generally drank too much, and made his presence unwelcome with the female ghosts as he decided that once he'd fucked the nano-droplets in one configuration, he might as well try fucking them in another. What the hell was it, Gus wondered, that made the living so reluctant to give up the dead, and the dead to give up the living?

In the few hours that they did spend together and alone at that time, Elanore and Gustav made detailed plans to travel. The idea was that they (meaning Elanore, with all the credit she was accumulating) would commission a ship, a sailing ship, traditional in every respect apart from the fact that the sails would be huge power receptors driven directly by the moon, and the spars would be the frame of a reality engine. Together, they would get away from all of this, and sail across the foreal oceans, perhaps even as far as Tahiti. Admittedly, Gustav was intrigued by the idea of returning to the painter who by now seemed to be the initial wellspring of his creativity. He was certainly in a suitably grumpy and isolationist mood to head off, as the poverty-stricken and desperate Gauguin had once done, in search of inspiration in the South Seas, and ultimately to his death from the prolonged effects of syphilis. But they never actually discussed what Tahiti would be *like*. Of course, there would be no tourists there now—only ec-

centrics bothered to travel foreal these days. Gustav liked
to think, in fact, that there would be none of the tall ugly
buildings and the huge Coca-Cola signs that he'd once
seen in an old photograph of Tahiti's main town of Pa-
peete. There might—who knows?—not be any reality engines, even, squatting like spiders across the beaches and
jungle. With the understandable way that the birthrate was
now declining, there would be just a few natives left, living as they had once lived before Cook and Bligh and all
the rest—even Gauguin with his art and his myths and his
syphilis—had ruined it for them. That was how Gustav
wanted to leave Tahiti.

Winter came to their clifftop house. The guests de-
parted. The wind raised white crests across the ocean. Gustav developed a habit, which Elanore pretended not to
notice, of turning the heating down; as if he needed chill
and discomfort to make the place seem real. Tahiti, that
ship of theirs, remained an impossibly long way off. There
were no final showdowns—just this gradual drifting apart.
Gustav gave up trying to make love to Elanore, just as he
had given up trying to paint her. But they were friendly and
cordial with each other. It seemed that neither of them
wished to pollute the memory of something that had once
been wonderful. Elanore was, Gustav knew, starting to become concerned about his failure to have his increasing
signs of age treated, and his refusal to have a librarian;
even his insistence on pursuing a career that seemed only
to leave him depleted and damaged. But she never said
anything.

They agreed to separate for a while. Elanore would
head off to explore pure virtuality. Gustav would go back
to foreal Paris and try to rediscover his art. And so, making promises they both knew they would never keep, Gustav and Elanore finally parted.

•   •   •

Gustav slid his unfinished *Olympia* back down amid the other canvases. He looked out of the window, and saw from the glow coming up through the gaps in the houses that the big reality engines were humming. The evening, or whatever other time and era it was, was in full swing.

A vague idea forming in his head, Gustav pulled on his coat and headed out from his tenement. As he walked down through the misty, smoggy streets, it almost began to feel like inspiration. Such was his absorption that he didn't even bother to avoid the shining bubbles of the reality engines. Paris, at the end of the day, still being Paris, the realities he passed through mostly consisted of one or another sort of café, but they were set amid dazzling souks, dank medieval alleys, yellow and seemingly watery places where swam strange creatures that he couldn't think to name. But his attention wasn't on it anyway.

The Musée D'Orsay was still kept in reasonably immaculate condition beside the faintly luminous and milky Seine. Outside and in, it was well-lit, and a trembling barrier kept in the air that was necessary to preserve its contents until the time came when they were fashionable again. Inside, it even *smelled* like an art gallery, and Gustav's footsteps echoed on the polished floors, and the robot janitors greeted him; in every way, and despite all the years since he'd last visited, the place was the same.

Gustav walked briskly past the statues and the bronze casts, past Ingres' big, dead canvases of supposedly voluptuous nudes. Then Moreau, early Degas, Corot, Millet . . . Gustav did his best to ignore them all. For the fact was that Gustav hated art galleries—he was still, at least, a painter in that respect. Even in the years when he'd gone deliberately to such places, because he knew that they were good for his own development, he still liked to think of himself as a kind of burglar—get in, grab your ideas, get out again. Everything else, all the ahhs and the oohs, was for mere spectators. . . .

He took the stairs to the upper floor. A cramp had worked its way beneath his diaphragm and his throat felt raw, but behind all of that there was this feeling, a tingling of power and magic and anger—a sense that perhaps . . .

Now that he was up amid the rooms and corridors of the great Impressionist works, he forced himself to slow down. The big gilt frames, the pompous marble, the names and dates of artists who had often died in anonymity, despair, disease, blindness, exile, near-starvation. Poor old Sisley's *Misty Morning.* Vincent Van Gogh in a self-portrait formed from deep, sensuous, three-dimensional oils. Genuinely great art was, Gustav thought, pretty depressing for would-be great artists. If it hadn't been for the invisible fields that were protecting these paintings, he would have considered ripping the things off the walls, destroying them.

His feet led him back to the Manets, that woman gazing out at him from *Dejéuner sur l'Herbe* and than again from *Olympia.* She wasn't beautiful, didn't even look much like Elanore. . . . But that wasn't the point. He drifted on past the clamoring canvases, wondering if the world had ever been this bright, this new, this wondrously chaotic. Eventually, he found himself face-to-face with the surprisingly few Gauguins that the Musée D'Orsay possessed. Those bright slabs of color, those mournful Tahitian natives, which were often painted on raw sacking because it was all Gauguin could get his hands on in the hot stench of his tropical hut. He became wildly fashionable after his death, of course; the idea of destitution on a faraway isle suddenly struck everyone as romantic. But it was too late for Gauguin by then. And too late—as his hitherto worthless paintings were snapped up by Russians, Danes, Englishmen, Americans—for these stupid, habitually arrogant Parisians. Gauguin was often poor at dealing with his shapes, but he generally got away with it. And his sense of color was like no one else's. Gustav remembered vaguely now that there was a nude that Gauguin had painted as his own lopsided tribute to Manet's *Olympia*—had even pinned a photograph of it to the wall of his hut as he worked. But, like most of Gauguin's other really important paintings, it wasn't here at the Musée D'Orsay, this supposed epicenter of Impressionist and Symbolist art. Gustav shrugged and turned away. He hobbled slowly back down through the galley.

Outside, beneath the moonlight, amid the nanosmog and the buzzing of the powerfield, Gustav made his way once again through the realities. An English teahouse circa 1930. A Guermantes salon. If they'd been foreal, he'd have sent the cups and the plates flying, bellowed in the self-satisfied faces of the dead and living. Then he stumbled into a scene he recognized from the Musée D'Orsay, one, in fact, that had once been as much a cultural icon as Madonna's tits or a Beatles tune. *Le Moulin de la Gallette.* He was surprised and almost encouraged to see Renoir's Parisian figures in their Sunday-best clothing dancing under the trees in the dappled sunlight, or chatting at the surrounding benches and tables. He stood and watched, nearly smiling. Glancing down, he saw that he was dressed appropriately in a rough woolen navy suit. He studied the figures, admiring their animation, the clever and, yes, convincing way that, through some trick of reality, they were composed. . . . Then he realized that he recognized some of the faces, and that they had also recognized him. Before he could turn back, he was called to and beckoned over.

"Gustav," Marcel's ghost said, sliding an arm around him, smelling of male sweat and Pernod. "Grab a chair. Sit down. Long time no see, eh?"

Gustav shrugged and accepted the brimming tumbler of wine that he offered. If it was foreal—which he doubted—this and a few more of the same might help him sleep tonight. "I thought you were in Venice," he said. "As the Doge."

Marcel shrugged. There were breadcrumbs on his mustache. "That was *ages* ago. Where have you been, Gustav?"

"Just around the corner, actually."

"Not still *painting,* are you?"

Gustav allowed that question to be lost in the music and the conversation's ebb and flow. He gulped his wine and looked around, expecting to see Elanore at any moment. So many of the others were here—it was almost like old times. There, even, was Francine, dancing with a top-hatted man—so she clearly wasn't across the sky. Gustav decided to ask the girl in the striped dress who was nearest to him if she'd seen Elanore. He realized as he spoke to her that her face was familiar to him, but he somehow couldn't recollect her name—even whether she was living or a ghost. She shook her head, and asked the woman who stood leaning behind her. But she, also, hadn't seen Elanore; not, at least, since the times when Marcel was in Venice and when Francine was across the sky. From there, the question rippled out across the square. But no one, it seemed, knew what had happened to Elanore.

Gustav stood up and made his way between the twirling dancers and the lantern-strung trees. His skin tingled as he stepped out of the reality, and the laughter and the music suddenly faded. Avoiding any other such encounters, he made his way back up the dim streets to his tenement.

There, back at home, the light from the setting moon was bright enough for him to make his way through the dim wreckage of his life without falling—and the terminal that Elanore's ghost had reactivated still gave off a virtual glow. Swaying, breathless, Gustav paged down into his accounts, and saw the huge sum—the kind of figure that he associated with astronomy, with the distance of the moon from the earth, the earth from the sun—that now appeared there. Then, he passed back through the terminal's levels, and began to search for Elanore.

But Elanore wasn't there.

•   •   •

Gustav was painting. When he felt like this, he loved and
hated the canvas in almost equal measures. The outside
world, foreal or in reality, ceased to exist for him.

A woman, naked, languid, and with a dusky skin quite
unlike Elanore's, is lying upon a couch, half-turned, her
face cupped in her hand that lies upon the primrose pillow,
her eyes gazing away from the onlooker at something far
off. She seems beautiful but unerotic, vulnerable yet
clearly available, and self-absorbed. Behind her—amid the
twirls of bright yet gloomy decoration—lies a glimpse of
stylized rocks under a strange sky, whilst two oddly dis-
turbing figures are talking, and a dark bird perches on the
lip of a balcony; perhaps a raven. . . .

Although he detests plagiarism, and is working solely
from memory, Gustav finds it hard to break away from
Gauguin's nude on this canvas he is now painting. But he
really isn't fighting that hard to do so, anyway. In this
above all of Gauguin's great paintings, stripped of the crap
and the despair and the self-justifying symbolism, Gauguin
was simply *right*. So Gustav still keeps working, and the
paint sometimes almost seems to want to obey him. He
doesn't know or care at the moment what the thing will
turn out like. If it's good, he might think of it as his tribute
to Elanore; and if it isn't . . . well, he knows that, once he's
finished this painting, he will start another one. Right now,
that's all that matters.

Elanore was right, Gustav decides, when she once said
that he was entirely selfish, and would sacrifice everything—himself included—just so that he could continue to
paint. She was eternally right and, in her own way, she too
was always searching for the next challenge, the next river
to cross. Of course, they should have made more of the
time that they had together, but as Elanore's ghost admitted at that Van Gogh café when she finally came to say
good-bye, nothing could ever quite be the same.

Gustav stepped back from his canvas and studied it,

eyes half-closed at first just to get the shape, then with a more appraising gaze. Yes, he told himself, and reminded himself to tell himself again later when he began to feel sick and miserable about it, this is a true work. This is worthwhile.

Then, and although there was much that he still had to do, and the oils were still wet, and he knew that he should rest the canvas, he swirled his brush in a blackish puddle of palette-mud and daubed the word NEVERMORE across the top, and stepped back again, wondering what to paint *next*.

APPROACHING PERIMELASMA

Geoffrey A. Landis

A physicist who works for NASA, and who has recently been working on the Martian Lander program, Geoffrey A. Landis is a frequent contributor to Analog *and to* Asimov's Science Fiction, *and has also sold stories to markets such as* Interzone, Amazing, *and* Pulphouse. *Landis is not a prolific writer, by the high-production standards of the genre, but he is popular. His story "A Walk in the Sun" won him a Nebula and a Hugo Award in 1992, his story "Ripples in the Dirac Sea" won him a Nebula Award in 1990, and his story "Elemental" was on the Final Huge Ballot a few years back. His first book was the collection,* Myths, Legends, and True History, *and he has just published his first novel,* Mars Crossing. *He lives in Brook Park, Ohio.*

Here he takes us along on a suspenseful and hair-raising cosmic ride that could only be made by someone who has moved far beyond flesh, as we accompany an intrepid (although nonliving, as we consider life from our provincial current perspective) future adventurer bound for someplace nobody has ever gone before: a headlong plunge into a black hole, and out of it again—if he can figure a way to get out of it, that is, with all the forces of the universe against him. . . .

There is a sudden frisson of adrenaline, a surge of something approaching terror (if I could still feel terror), and I realize that this is it, this time I am the one who is doing it.

I'm the one who is going to drop into a black hole.
Oh, my God. This time I'm not you.
This is real.
Of course, I have experienced this exact feeling before.
We both know exactly what it feels like.

My body seems weird, too big and at once too small. The
feel of my muscles, my vision, my kinesthetic sense,
everything is wrong. Everything is strange. My vision is
fuzzy, and colors are oddly distorted. When I move, my
body moves unexpectedly fast. But there seems to be nothing wrong with it. Already I am getting used to it. "It will
do," I say.

There is too much to know, too much to be all at once.
I slowly coalesce the fragments of your personality. None
of them are you. All of them are you.

A pilot, of course, you must have, you must be, a pilot.
I integrate your pilot persona, and he is me. I will fly to the
heart of a darkness far darker than any mere unexplored
continent. A scientist, somebody to understand your experience, yes. I synthesize a persona. You are him, too, and I
understand.

And someone to simply *experience* it, to tell the tale (if
any of me will survive to tell the tale) of how you dropped
into a black hole, and how you survived. If you survive.
Me. I will call myself Wolf, naming myself after a nearby
star, for no reason whatsoever, except maybe to claim, if
only to myself, that I am not you.

All of we are me are you. But in a real sense, you're not
here at all. None of me are you. You are far away. Safe.

Some black holes, my scientist persona whispers, are deco-
rated with an accretion disk, shining like a gaudy signal
in the sky. Dust and gas from the interstellar medium fall

toward the hungry singularity, accelerating to nearly the speed of light in their descent, swirling madly as they fall. It collides; compresses; ionizes. Friction heats the plasma millions of degrees, to emit a brilliant glow of hard X rays. Such black holes are anything but black; the incandescence of the infalling gas may be the most brilliantly glowing thing in a galaxy. Nobody and nothing would be able to get near it; nothing would be able to survive the radiation.

The Virgo hole is not one of these. It is ancient, dating from the very first burst of star-formation when the universe was new, and has long ago swallowed or ejected all the interstellar gas in its region, carving an emptiness far into the interstellar medium around it.

The black hole is fifty-seven light years from Earth. Ten billion years ago, it had been a supermassive star, and exploded in a supernova that for a brief moment had shone brighter than the galaxy, in the process tossing away half its mass. Now there is nothing left of the star. The burned-out remnant, some thirty times the mass of the sun, has pulled in space itself around it, leaving nothing behind but gravity.

Before the download, the psychologist investigated my—your—mental soundness. We must have passed the test, obviously, since I'm here. What type of man would allow himself to fall into a black hole? That is my question; maybe if I can answer that, I would understand ourself.

But this did not seem to interest the psychologist. She did not, in fact, even look directly at me. Her face had the focusless abstract gaze characteristic of somebody hotlinked by the optic nerve to a computer system. Her talk was perfunctory. To be fair, the object of her study was not the flesh me, but my computed reflection, the digital maps of my soul. I remember the last thing she said.

"We are fascinated with black holes because of their depth of metaphor," she said, looking nowhere. "A black hole is, literally, the place of no return. We see it as a metaphor for how we, ourselves, are hurled blindly into a place from which no information ever reaches us, the place from which no one ever returns. We live our lives falling into the future, and we will all inevitably meet the singularity." She paused, expecting, no doubt, some comment. But I remained silent.

"Just remember this," she said, and for the first time her eyes returned to the outside world and focused on me. "This is a real black hole, not a metaphor. Don't treat it like a metaphor. Expect reality." She paused, and finally added, "Trust the math. It's all we really know, and all that we have to trust."

Little help.

Wolf versus the black hole! One might think that such a contest is an unequal one, that the black hole has an overwhelming advantage.

Not quite so unequal.

On my side, I have technology. To start with, the wormhole, the technological sleight-of-space which got you fifty-seven light-years from Earth in the first place.

The wormhole is a monster of relativity no less than the black hole, a trick of curved space allowed by the theory of general relativity. After the Virgo black hole was discovered, a wormhole mouth was laboriously dragged to it, slower than light, a project that took over a century. Once the wormhole was here, though, the trip became only a short one, barely a meter of travel. Anybody could come here and drop into it.

A wormhole—a far too cute name, but one we seem to be stuck with—is a shortcut from one place to another. Physically, it is nothing more than a loop of exotic matter.

If you move through the hoop on this side of the wormhole, you emerge out the hoop on that side. Topologically,
the two sides of the wormhole are pasted together, a piece
cut out of space glued together elsewhere.

Exhibiting an excessive sense of caution, the proctors
of Earthspace refused to allow the other end of the Virgo
wormhole to exit at the usual transportation nexus, the
wormhole swarm at Neptune-Trojan 4. The far end of
the wormhole opens instead to an orbit around Wolf-562,
an undistinguished red dwarf sun circled by two airless
planets that are little more than frozen rocks, twenty-one
light-years from Earthspace. To get here we had to take a
double wormhole hop: Wolf, Virgo.

The black hole is a hundred kilometers across. The
wormhole is only a few meters across. I would think that
they were overly cautious.

The first lesson of relativity is that time and space are
one. For a long time after the theoretical prediction that
such a thing as a traversable wormhole ought to be possible, it was believed that a wormhole could also be made to
traverse time as well. It was only much later, when wormhole travel was tested, that it was found that the Cauchy instability makes it impossible to form a wormhole that leads
backward in time. The theory was correct—space and time
are indeed just aspects of the same reality, spacetime—but
any attempt to move a wormhole in such a way that it
becomes a timehole produces a vacuum polarization to
cancel out the time effect.

After we—the spaceship I am to pilot, and myself/your-
self—come through the wormhole, the wormhole engineers go to work. I have never seen this process close up,
so I stay nearby to watch. This is going to be interesting.

A wormhole looks like nothing more than a circular
loop of string. It is, in fact, a loop of exotic material,
negative-mass cosmic string. The engineers, working tele-
robotically via vacuum manipulator pods, spray charge

onto the string. They charge it until it literally glows with Paschen discharge, like a neon light in the dirty vacuum, and then use the electric charge to manipulate the shape. With the application of invisible electromagnetic fields, the string starts to twist. This is a slow process. Only a few meters across, the wormhole loop has a mass roughly equal to that of Jupiter. Negative to that of Jupiter, to be precise, my scientist persona reminds me, but either way, it is a slow thing to move.

Ponderously, then, it twists further and further, until at last it becomes a lemniscate, a figure of eight. The instant the string touches itself, it shimmers for a moment, and then suddenly there are two glowing circles before us, twisting and oscillating in shape like jellyfish.

The engineers spray more charge onto the two wormholes, and the two wormholes, arcing lightning into space, slowly repel each other. The vibrations of the cosmic string are spraying our gravitational radiation like a dog shaking off water—even where I am, floating ten kilometers distant, I can feel it, like the swaying of invisible tides—and as they radiate energy, the loops enlarge. The radiation represents a serious danger. If the engineers lose control of the string for even a brief instant, it might enter the instability known as "squiggle mode," and catastrophically enlarge. The engineers damp out the radiation before it gets critical, though—they are, after all, well practiced at this—and the loops stabilize into two perfect circles. On the other side, at Wolf, precisely the same scene has played out, and two loops of exotic string now circle Wolf-562 as well. The wormhole has been cloned.

All wormholes are daughters of the original wormhole, found floating in the depths of interstellar space eleven hundred years ago, a natural loop of negative cosmic string as ancient as the Big Bang, invisible to the eyes save for the distortion of spacetime. That first one led from nowhere interesting to nowhere exciting, but from that one we bred hundreds, and now we casually move wormhole mouths from star to star, breeding new wormholes as it suits us, to form an ever-expanding network of connections.

I should not have been so close. Angry red lights have been flashing in my peripheral vision, warning blinkers that I have been ignoring. The energy radiated in the form of gravitational waves had been prodigious, and would have, to a lesser person, been dangerous. But in my new body, I am nearly invulnerable, and if I can't stand a mere wormhole cloning, there is no way I will be able to stand a black hole. So I ignore the warnings, wave briefly to the engineers—though I doubt that they can even see me, floating kilometers away—and use my reaction jets to scoot over to my ship.

The ship I will pilot is docked to the research station, where the scientists have their instruments and the biological humans have their living quarters. The wormhole station is huge compared to my ship, which is a tiny ovoid occupying a berth almost invisible against the hull. There is no hurry for me to get to it.

I'm surprised that any of the technicians can even see me, tiny as I am in the void, but a few of them apparently do, because in my radio I hear casual greetings called out: How's it, *ohayo gozaimasu,* hey glad you made it, how's the bod? It's hard to tell from the radio voices which ones are people I know, and which are only casual acquaintances. I answer back: How's it, *ohayo,* yo, surpassing spec. None of them seem inclined to chat, but then, they're busy with their own work.

They are dropping things into the black hole.

Throwing things in, more to say. The wormhole station orbits a tenth of an astronomical unit from the Virgo black hole, closer to the black hole than Mercury is to the sun.

This is an orbit with a period of a little over two days, but, even so close to the black hole, there is nothing to see. A rock, released to fall straight downward, takes almost a day to reach the horizon.

One of the scientists supervising, a biological human named Sue, takes the time to talk with me a bit, explaining what they are measuring. What interests me most is that they are measuring whether the fall deviates from a straight line. This will let them know whether the black hole is rotating. Even a slight rotation would mess up the intricate dance of the trajectory required for my ship. However, the best current theories predict that an old black hole will have shed its angular momentum long ago, and, as far as the technicians can determine, their results show that the conjecture holds.

The black hole, or the absence in space where it is located, is utterly invisible from here. I follow the pointing finger of the scientist, but there is nothing to see. Even if I had a telescope, it is unlikely that I would be able to pick out the tiny region of utter blackness against the irregular darkness of an unfamiliar sky.

My ship is not so different from the drop probes. The main difference is that I will be on it.

Before boarding the station, I jet over in close to inspect my ship, a miniature egg of perfectly reflective material. The hull is made of a single crystal of a synthetic material so strong that no earthly force could even dent it.

A black hole, though, is no earthly force.

Wolf versus the black hole! The second technological trick I have in my duel against the black hole is my body.

I am no longer a fragile, fluid-filled biological human. The tidal forces at the horizon of a black hole would rip a true human apart in mere instants; the accelerations required to hover would squash one into liquid. To make this journey, I have downloaded your fragile biological mind into a body of more robust material. As important as the strength of my new body is the fact that it is tiny. The force produced by the curvature of gravity is proportional to the size of the object. My new body, a millimeter tall, is millions of times more resistant to being stretched to spaghetti.

The new body has another advantage as well. With my mind operating as software on a computer the size of a pinpoint, my thinking and my reflexes are thousands of times faster than biological. In fact, I have already chosen to slow my thinking down, so that I can still interact with the biologicals. At full speed, my microsecond reactions are lightning compared to the molasses of neuron speeds in biological humans. I see far in the ultraviolet now, a necessary compensation for the fact that my vision would consist of nothing but a blur if I tried to see by visible light.

You could have made my body any shape, of course, a tiny cube or even a featureless sphere. But you followed the dictates of social convention. A right human should be recognizably a human, even if I am to be smaller than an ant, and so my body mimics a human body, although no part of it is organic, and my brain faithfully executes your own human brain software. From what I see and feel, externally and internally, I am completely, perfectly human.

As is right and proper. What is the value of experience to a machine?

Later, after I return—*if* I return—I can upload back. I can become you.

But return is, as they say, still somewhat problematical.

You, my original, what do you feel? Why did I think I would do it? I imagine you laughing hysterically about the trick you've played, sending me to drop into the black hole while you sit back in perfect comfort, in no danger. Imagining your laughter comforts me, for all that I know that it is false. I've been in the other place before, and never laughed.

I remember the first time I fell into a star.

We were hotlinked together, that time, united in online-realtime, our separate brains reacting as one brain. I remember what I thought, the incredible electric feel: OhmiGod, am I really going to do this? Is it too late to back out?

The idea had been nothing more than a whim, a crazy idea, at first. We had been dropping probes into a star, Groombridge 1830B, studying the dynamics of a flare star. We were done, just about, and the last-day-of-project party was just getting in swing. We were all fuzzed with neuro-transmitter randomizers, creativity spinning wild and critical thinking nearly zeroes. Somebody, I think it was Jenna, said we could ride one down, you know. Wait for a flare, and then plunge through the middle of it. Helluva ride!

Helluva *splash* at the end, too, somebody said, and laughed.

Sure, somebody said. It might have been me. What do you figure? Download yourself to temp storage and then uplink frames from yourself as you drop?

That works, Jenna said. Better: We copy our bodies first, then link the two brains. One body drops; the other copy hotlinks to it.

Somehow, I don't remember when, the word *we* had grown to include me.

"Sure," I said. "And the copy on top is in null-input suspension; experiences the whole thing realtime!"

In the morning, when we were focused again, I might have dismissed the idea as a whim of the fuzz, but for

Jenna the decision was already immovable as a droplet of neutronium. *Sure* we're dropping, let's start now.

We made a few changes. It takes a long time to fall into a star, even a small one like Bee, so the copy was reengineered to a slower thought-rate, and the original body in null-input was frame-synched to the drop copy with impulse-echoers. Since the two brains were molecule by molecule identical, the uplink bandwidth required was minimal.

The probes were reworked to take a biological, which meant mostly that a cooling system had to be added to hold the interior temperature within the liquidus range of water. We did that by the simplest method possible: We surrounded the probes with a huge block of cometary ice. As it sublimated, the ionized gas would carry away heat. A secondary advantage of the ice was that our friends, watching from orbit, would have a blazing cometary trail to cheer on. When the ice was used up, of course, the body would slowly vaporize. None of us would actually survive to hit the star.

But that was no particular concern. If the experience turned out to be too undesirable, we could always edit the pain part of it out of the memory later.

It would have made more sense, perhaps, to have simply recorded the brain-uplink from the copy onto a local high-temp buffer, squirted it back, and linked it to as a memory upload. But Jenna would have none of that. She wanted to experience it in realtime, or at least in as close to realtime as speed-of-light delays allow.

Three of us—Jenna, Martha, and me—dropped. Something seems to be missing from my memory here; I can't remember the reason I decided to do it. It must have been something about a biological body, some arational consideration that seemed normal to my then-body, that I could never back down from a crazy whim of Jenna's.

And I had the same experience, the same feeling then,

as I, you, did, always do, the feeling that my God *I* am the copy and I am going to die. But that time, of course, thinking every thought in synchrony, there was no way at all to tell the copy from the original, to split the me from you.

It is, in its way, a glorious feeling.

I dropped.

You felt it, you remember it. Boring at first, the long drop with nothing but freefall and the chatter of friends over the radio-link. Then the ice shell slowly flaking away, ionizing and beginning to glow, a diaphanous cocoon of pale violet, and below the red star getting larger and larger, the surface mottled and wrinkled, and then suddenly we fell into and through the flare, a huge luminous vault above us, dwarfing our bodies in the immensity of creation.

An unguessable distance beneath me, the curvature of the star vanished, and, still falling at three hundred kilometers per second, I was hanging motionless over an infinite plane stretching from horizon to horizon.

And then the last of the ice vaporized, and I was suddenly suspended in nothing, hanging nailed to the burning sky over endless crimson horizons of infinity, and pain came like the inevitability of mountains—I didn't edit it—pain like infinite oceans, like continents, like a vast, airless world.

Jenna, now I remember. The odd thing is, I never did really connect in any significant way with Jenna. She was already in a quadrad of her own, a quadrad she was fiercely loyal to, one that was solid and accepting to her chameleon character, neither needing nor wanting a fifth for completion.

Long after, maybe a century or two later, I found out that Jenna had disassembled herself. After her quadrad split apart, she'd downloaded her character to a mainframe, and then painstakingly cataloged everything that made her Jenna: all her various skills and insights, everything she had experienced, no matter how minor, each facet of her character, every memory and dream and longing: the myriad subroutines of personality. She indexed her soul, and she put the ten thousand pieces of it into the public domain for download. A thousand people, maybe a million people, maybe even more have pieces of Jenna, her cleverness, her insight, her skill at playing antique instruments.

But nobody has her sense of self. After she copied her subroutines, she deleted herself.

And who am I?

Two of the technicians who fit me into my spaceship and who assist in the ten thousand elements of the preflight check are the same friends from that drop, long ago; one of them even still in the same biological body as he had then, although eight hundred years older, his vigor undiminished by biological reconstruction. My survival, if I am to survive, will be dependent on microsecond timing, and I'm embarrassed not to be able to remember his name.

He was, I recall, rather stodgy and conservative even back then.

We joke and trade small talk as the checkout proceeds. I'm still distracted by my self-questioning, the implications of my growing realization that I have no understanding of why I'm doing this.

Exploring a black hole would be no adventure if only we had faster-than-light travel, but of the thousand technological miracles of the third and fourth millennia, this one miracle was never realized. If I had the mythical FTL motor, I could simply drive out of the black hole. At the event horizon, space falls into the black hole at the speed of light; the mythical motor would make that no barrier.

But such a motor we do not have. One of the reasons I'm

taking the plunge—not the only one, not the main one, but one—is in the hope that scientific measurements of the warped space inside the black hole will elucidate the nature of space and time, and so I myself will make one of the innumerable small steps to bring us closer to an FTL drive.

The spaceship I am to pilot has a drive nearly—but not quite—as good. It contains a microscopic twist of space-time inside an impervious housing, a twist that will parity-reverse ordinary matter into mirror-matter. This total conversion engine gives my ship truly ferocious levels of thrust. The gentlest nudge of my steering rockets will give me thousands of gravities of acceleration. Unthinkable acceleration for a biological body, no matter how well cushioned. The engine will allow the rocket to dare the unthinkable, to hover at the very edge of the event horizon, to maneuver where space itself is accelerating at nearly light-speed. This vehicle, no larger than a peanut, contains the engines of an interstellar probe.

Even with such an engine, most of the ship is reaction mass.

The preflight checks are all green. I am ready to go. I power up my instruments, check everything out for myself, verify what has already been checked three times, and then check once again. My pilot persona is very thorough. Green.

"You still haven't named your ship," comes a voice to me. It is the technician, the one whose name I have forgotten. "What is your call sign?"

One way journey, I think. Maybe something from Dante? No, Sartre said it better: no exit. "*Huis Clos*," I say, and drop free.

Let them look it up.

Alone.
The laws of orbital mechanics have not been sus-

pended, and I do not drop into the black hole. Not yet. With the slightest touch of my steering engines—I do not dare use the main engine this close to the station—I drop into an elliptical orbit, one with a perimelasma closer to, but still well outside, the dangerous zone of the black hole. The black hole is still invisible, but inside my tiny kingdom I have enhanced senses of exquisite sensitivity, spreading across the entire spectrum from radio to gamma radiation. I look with my new eyes to see if I can detect an X-ray glow of interstellar hydrogen being ripped apart, but if there is any such, it is too faint to be visible with even my sensitive instruments. The interstellar medium is so thin here as to be essentially nonexistent. The black hole is invisible.

I smile. This makes it better, somehow. The black hole is pure, unsullied by any outside matter. It consists of gravity and nothing else, as close to pure mathematical abstraction as anything in the universe can ever be.

It is not too late to back away. If I were to choose to accelerate at a million gravities, I would reach relativistic velocities in about thirty seconds. No wormholes would be needed for me to run away; I would barely even need to slow down my brain to cruise at nearly the speed of light to anywhere in the colonized galaxy.

But I know I won't. The psychologist knew it, too, damn her, or she would never have approved me for the mission. Why? What is it about me?

As I worry about this with part of my attention, while the pilot persona flies the ship, I flash onto a realization, and at this realization another memory hits. It is the psychologist, and in the memory I'm attracted to her sexually, so much so that you are distracted from what she is saying.

I feel no sexual attraction now, of course. I can barely remember what it is. That part of the memory is odd, alien.

"We can't copy the whole brain to the simulation, but we can copy enough that, to yourself, you will still feel like yourself," she said. She is talking to the air, not to you. "You won't notice any gaps."

I'm brain-damaged. This is the explanation.

You frowned. "How could I not notice that some of my memories are missing?"

"The brain makes adjustments. Remember, at any given time, you never even use 1 percent of 1 percent of your memories. What we'll be leaving out will be stuff that you will never have any reason to think about. The memory of the taste of strawberries, for example; the floor-plan of the house you lived in as a teenager. Your first kiss."

This bothered you somewhat—you want to remain yourself. I concentrate, hard. What do strawberries taste like? I can't remember. I'm not even certain what color they are. Round fruits, like apples, I think, only smaller. And the same color as apples, or something similar, I'm sure, except I don't remember what color that is.

You decided that you can live with the editing, as long as it doesn't change the essential you. You smiled. "Leave in the first kiss."

So I can never possibly solve the riddle: What kind of a man is it that would deliberately allow himself to drop into a black hole. I cannot, because I don't have the memories of you. In a real sense, I am not you at all.

But I do remember the kiss. The walk in the darkness, the grass wet with dew, the moon a silver sliver on the horizon, turning to her, and her face already turned up to meet my lips. The taste indescribable, more feeling than taste (not like strawberries at all), the small hardness of her teeth behind the lips—all there. Except the one critical detail: I don't have any idea at all who she *was*.

What else am I missing? Do I even know what I don't know?

I was a child, maybe nine, and there was no tree in the neighborhood that you could not climb. I was a careful, meticulous, methodical climber. On the tallest of the trees, when you reached toward the top, you were above the forest canopy (did I live in a forest?) and, out of the dimness of the forest floor, emerged into brilliant sunshine. Nobody else could climb like you; nobody ever suspected how high I climbed. It was your private hiding place, so high that the world was nothing but a sea of green waves in the valley between the mountains.

It was my own stupidity, really. At the very limit of the altitude needed to emerge into sunlight, the branches were skinny, narrow as your little finger. They bent alarmingly with your weight, but I knew exactly how much they would take. The bending was a thrill, but I was cautious, and knew exactly what I was doing.

It was farther down, where the branches were thick and safe, that I got careless. Three points of support, that was the rule of safety; but I was reaching for one branch, not paying attention, when one in my other hand broke, and I was off balance. I slipped. For a prolonged instant I was suspended in space, branches all about me, I reached out and grasped only leaves, and I fell and fell, and all I could think as leaves and branches fell upward past me was, oh my, I made a miscalculation; I was really stupid.

The flash memory ends with no conclusion. I must have hit the ground, but I cannot remember it. Somebody must have found me, or else I wandered or crawled back, perhaps in a daze, and found somebody, but I cannot remember it.

Half a million kilometers from the hole. If my elliptical orbit were around the sun instead of a black hole, I would already have penetrated the surface. I now hold the record for the closest human approach. There is still nothing to see with unmagnified senses. It seems surreal that I'm in the grip of something so powerful that it is utterly invisible. With my augmented eyes used as a telescope, I can detect the black hole by what isn't there, a tiny place of blackness nearly indistinguishable from any other patch of darkness except for an odd motion of the stars near it.

My ship is sending a continuous stream of telemetry back to the station. I have an urge to add a verbal commentary—there is plenty of bandwidth—but I have nothing to say. There is only one person I have any interest in talking to, and you are cocooned at absolute zero, waiting for me to upload myself and become you.

My ellipse takes me inward, moving faster and faster. I am still in Newton's grip, far from the sphere where Einstein takes hold.

A tenth of a solar radius. The blackness I orbit is now large enough to see without a telescope, as large as the sun seen from Earth, and swells as I watch with time-distorted senses. Due to its gravity, the blackness in front of the star pattern is a bit larger than the disk of the black hole itself. Square root of twenty-seven over two—about two and a half times larger, the physicist persona notes. I watch in fascination.

What I see is a bubble of purest blackness. The bubble pushes the distant stars away from it as it swells. My orbital motion makes the background stars appear to sweep across the sky, and I watch them approach the black hole and then, smoothly pushed by the gravity, move off to the side, a river of stars flowing past an invisible obstacle. It is a gravitational lensing effect, I know, but the view of flowing stars is so spectacular that I cannot help but watch it. The gravity pushes each star to one side or the other. If a star were to pass directly behind the hole, it would appear to split and for an instant become a perfect circle of light, an Einstein ring. But this precise alignment is too rare to see by accident.

Closer, I notice an even odder effect. The sweeping stars detour smoothly around the bubble of blackness, but very close to the bubble, there are other stars, stars that actually move in the opposite direction, a counterflowing river of stars. It takes me a long time (microseconds perhaps) before my physicist persona tells me that I am seeing the image of the stars in the Einstein mirror. The entire external universe is mirrored in a narrow ring outside the black hole, and the mirror image flows along with a mirror of my own motion.

In the center of the ring there is nothing at all.

Five thousand kilometers, and I am moving fast. The gravitational acceleration here is over ten million gees, and I am still fifty times the Schwarzchild radius from the black hole. Einstein's correction is still tiny, though, and if I were to do nothing, my orbit would whip around the black hole and still escape into the outside world.

One thousand kilometers. Perimelasma, the closest point of my elliptical orbit. Ten times the Schwarzchild radius, close enough that Einstein's correction to Newton now makes a small difference to the geometry of space. I fire my engines. My speed is so tremendous that it takes over a second of my engine firing at a million gravities to circularize my orbit.

My time sense has long since speeded up back to normal, and then faster than normal. I orbit the black hole about ten times per second.

My god, this is why I exist, this is why I'm here!

All my doubts are gone in the rush of naked power. No biological could have survived this far; no biological could have even survived the million-gee circularization burn, and I am only at the very beginning! I grin like a maniac, throb with a most unscientific excitement that must be the electronic equivalent of an adrenaline high.

Oh, this ship is good. This ship is sweet. A million-gee burn, smooth as magnetic levitation, and I barely cracked the throttle. I should have taken it for a spin before dropping in, should have hot-rodded *Huis Clos* around the stellar neighborhood. But it had been absolutely out of the question to fire the main engine close to the wormhole station. Even with the incredible efficiency of the engine, that million-gee perimelasma burn must have lit up the research station like an unexpected sun.

I can't wait to take *Huis Clos* in and see what it will *really* do.

My orbital velocity is a quarter of the speed of light.

The orbit at nine hundred kilometers is only a parking orbit, a chance for me to configure my equipment, make final measurements, and, in principle, a last chance for me to change my mind. There is nothing to reconnoiter that the probes have not already measured, though, and there is no chance that I will change my mind, however sensible that may seem.

The river of stars swirls in a dance of counterflow around the blackness below me. The horizon awaits.

The horizon below is invisible, but real. There is no barrier at the horizon, nothing to see, nothing to feel. I will even be unable to detect it, except for my calculations.

An event horizon is a one-way membrane, a place you can pass into but neither you nor your radio signals can pass out of. According to the mathematics, as I pass through the event horizon, the directions of space and time change identity. Space rotates into time; time rotates into space. What this means is that the direction to the center of the black hole, after I pass the event horizon, will be the future. The direction out of the black hole will be the past. This is the reason that no one and nothing can ever leave a black hole; the way inward is the one direction we always must go, whether we will it or not: into the future.

Or so the mathematics says.

The future, inside a black hole, is a very short one.

So far the mathematics has been right on. Nevertheless, I go on. With infinitesimal blasts from my engine, I inch my orbit lower.

The bubble of blackness gets larger, and the counterflow of stars around it becomes more complex. As I approach three times the Schwarzchild radius, 180 kilometers, I check all my systems. This is the point of no rescue: Inside three Schwarzchild radii, no orbits are stable, and my automatic systems will be constantly thrusting to adjust my orbital parameters to keep me from falling into the black hole or being flung away to infinity. My systems are all functional, in perfect form for the dangerous drop. My orbital velocity is already half the speed of light. Below this point, centrifugal force will decrease toward zero as I lower my orbit, and I must use my thrusters to increase my velocity as I descend, or else plunge into the hole.

When I grew up, in the last years of the second millennium, nobody thought that they would live forever. Nobody would have believed me if I told them that by my thousandth birthday, I would have no concept of truly dying.

Even if all our clever tricks fail, even if I plunge through the event horizon and am stretched into spaghetti and crushed by the singularity, I will not die. You, my original, will live on, and if *you* were to die, we have made dozens of backups and spin-off copies of myselves in the past, some versions of which must surely still be living on. My individual life has little importance. I can, if I chose, unplink my brain-state to the orbiting station right at this instant, and reawake, whole, continuing this exact thought, unaware (except on an abstract intellectual level) that I and you are not the same.

But we are not the same, you and I. I am an edited-down version of you, and the memories that have been edited out, even if I never happen to think them, make me different, a new individual. Not *you*.

On a metaphorical level, a black hole stands for death, the blackness that is sucking us all in. But what meaning does death have in a world of matrix backups, and modular personality? Is my plunge a death wish? Is it thumbing my nose at death? Because I intend to survive. Not you. *Me*.

I orbit the black hole over a hundred times a second now, but I have revved my brain processing speed accordingly, so that my orbit seems to me leisurely enough. The view here is odd. The black hole has swollen to the size of a small world below me, a world of perfect velvet darkness, surrounded by a belt of madly rotating stars.

No engine, no matter how energetic, can put a ship into an orbit at 1.5 times the Schwarzchild radius; at this distance, the orbital velocity is the speed of light, and not even my total-conversion engine can accelerate me to that speed. Below that, there are no orbits at all. I stop my descent at an orbit just sixty kilometers from the event horizon, when my orbital velocity reaches 85 percent of the speed of light. Here I can coast, ignoring the constant small adjustments of the thrusters that keep my orbit from sliding off the knife-edge. The velvet blackness of the black hole is almost half of the universe now, and if I were to trust the outside view, I am diving at a slant downward into the black hole. I ignore my pilot's urge to override the automated navigation and manually even out the trajectory. The downward slant is only relativist aberration, nothing more, an illusion of my velocity.

And 85 percent of the speed of light is as fast as I dare orbit; I must conserve my fuel for the difficult part of the plunge to come.

In my unsteady orbit sixty kilometers above the black hole, I let my ship's computer chat with the computer of the wormhole station, updating and downloading my sensors' observations.

At this point, according to the mission plan, I am supposed to uplink my brain state, so that should anything go wrong farther down the well, you, my original, will be able to download my state and experiences to this point. To hell with that, I think, a tiny bit of rebellion. I am not you. If you awaken with my memories, I will be no less dead.

Nobody at the wormhole station questions my decision not to upload.

I remember one other thing now. "You're a type N personality," the psychologist had said, twitching her thumb to leaf through invisible pages of test results. The gesture marked her era; only a person who had grown up before computer hotlinks would move a physical muscle in commanding a virtual. She was twenty-first century, possibly even twentieth. "But I suppose you already know that."

"Type N?" you asked.

"Novelty-seeking," she said. "Most particularly, one not prone to panic at new situations."

"Oh," you said. You did already know that. "Speaking of novelty seeking, how do you feel about going to bed with a type N personality?"

"That would be unprofessional." She frowned. "I think."

"Not even one who is about to jump down a black hole?"

She terminated the computer link with a flick of her wrist, and turned to look at you. "Well—"

From this point onward, microsecond timing is necessary for the dance we have planned to succeed. My computer and the station computer meticulously compare clocks, measuring Doppler shifts to exquisite precision. My clocks are running slow, as expected, but half of the slowness is relativistic time dilation due to my velocity. The gravitational redshift is still modest. After some milliseconds—a long wait for me, in my hyped-up state—they declare that they agree. The station has already done their part, and I begin the next phase of my descent.

The first thing I do is fire my engine to stop my orbit. I

crack the throttle to fifty million gees of acceleration, and
the burn takes nearly a second, a veritable eternity, to slow
my flight.

For a moment I hover, and start to drop. I dare not drop
too fast, and I ramp my throttle up, to a hundred megagee,
five hundred, a billion gravities. At forty billion gravities
of acceleration, my engine thrust equals the gravity of the
black hole, and I hover.

The blackness has now swallowed half of the universe.
Everything beneath me is black. Between the black below
and the starry sky above, a spectacularly bright line exactly
bisects the sky. I have reached the altitude at which orbital
velocity is just equal to the speed of light, and the light
from my rocket exhaust is in orbit around the black hole.
The line I see around the sky is my view of my own rocket,
seen by light that has traveled all the way around the black
hole. All I can see is the exhaust, far brighter than anything
else in the sky.

The second brightest thing is the laser beacon from the
wormhole station above me, shifted from the original red
laser color to a greenish blue. The laser marks the exact
line between the station and the black hole, and I maneuver carefully until I am directly beneath the orbiting station.

At forty billion gravities, even my ultrastrong body is at
its limits. I cannot move, and even my smallest finger is
pressed against the formfitting acceleration couch. But the
controls, hardware-interfaced to my brain, do not require
me to lift a finger to command the spacecraft. The command I give *Huis Clos* is: down.

My engine throttles down slightly, and I drop inward
from the photon sphere, the bright line of my exhaust vanishes. Every stray photon from my drive is now sucked
downward.

Now my view of the universe has changed. The black
hole has become the universe around me, and the universe

itself, all the galaxies and stars and the wormhole station, is a shrinking sphere of sparkling dust above me.

Sixty billion gravities. Seventy. Eighty.

Eighty billion gravities is full throttle. I am burning fuel at an incredible rate, and only barely hold steady. I am still twenty kilometers above the horizon.

There is an unbreakable law of physics: Incredible accelerations require incredible fuel consumption. Even though my spaceship is, by mass, comprised mostly of fuel, I can maintain less than a millisecond worth of thrust at this acceleration. I cut my engine and drop.

It will not be long now. This is my last chance to uplink a copy of my mind back to the wormhole station to wake in your body, with my last memory the decision to upload my mind.

I do not.

The stars are blueshifted by a factor of two, which does not make them noticeably bluer. Now that I have stopped accelerating, the starlight is falling into the hole along with me, and the stars do not blueshift any further. My instruments probe the vacuum around me. The theorists say that the vacuum close to the horizon of a black hole is an exotic vacuum, a bristle with secret energy. Only a ship plunging through the event horizon would be able to measure this. I do, recording the results carefully on my ship's onboard recorders, since it is now far too late to send anything back by radio.

There is no sign to mark the event horizon, and there is no indication at all when I cross it. If it were not for my computer, there would be no way for me to tell that I have passed the point of no return.

Nothing is different. I look around the tiny cabin, and can see no change. The blackness below me continues to grow, but is otherwise not changed. The outside universe continues to shrink above me; the brightness beginning to concentrate into a belt around the edge of the glowing sphere of stars, but this is only an effect of my motion. The only difference is that I have only a few hundred microseconds left.

From the viewpoint of the outside world, the light from my spacecraft has slowed down and stopped at the horizon. But I have far outstripped my lagging image, and am falling toward the center at incredible speed. At the exact center is the singularity, far smaller than an atom, a mathematical point of infinite gravity and infinite mystery.

Whoever I am, whether or not I survive, I am now the first person to penetrate the event horizon of a black hole. That's worth a cheer, even with nobody to hear. Now I have to count on the hope that the microsecond timing of the technicians above me had been perfect for the second part of my intricate dance, the part that might, if all goes well, allow me to survive.

Above me, according to theory, the stars have already burned out, and even the most miserly red dwarf has sputtered the last of its hydrogen fuel and grown cold. The universe has already ended, and the stars have gone out. I still see a steady glow of starlight from the universe above me, but this is fossil light, light that has been falling down into the black hole with me for eons, trapped in the infinitely stretched time of the black hole.

For me, time has rotated into space, and space into time. Nothing feels different to me, but I cannot avoid the singularity at the center of the black hole any more than I can avoid the future. Unless, that is, I have a trick.

Of course I have a trick.

At the center of the spherical universe above me is a dot of bright blue-violet, the fossil light of the laser beacon from the orbiting station. My reaction jets have kept on adjusting my trajectory to keep me centered in the guidance beam, so I am directly below the station. Anything dropped from the station will, if everything works right, drop directly onto the path I follow.

I am approaching close to the center now, and the tidal forces stretching my body are creeping swiftly toward a billion gees per millimeter. Much higher, and even my tremendously strong body will be ripped to spaghetti. There are only microseconds left for me. It is time.

I hammer my engine, full throttle. Far away, and long ago, my friends at the wormhole station above dropped a wormhole into the event horizon. If their timing was perfect—

From a universe that has already died, the wormhole cometh.

Even with my enhanced time sense, things happen fast. The laser beacon blinks out, and the wormhole sweeps down around me like the vengeance of God, far faster than I can react. The sparkle-filled sphere of the universe blinks out like a light, and the black hole—and the tidal forces stretching my body—abruptly disappears. For a single instant I see a black disk below me, and then the wormhole rotates, twists, stretches, and then silently vanishes.

Ripped apart by the black hole.

My ship is vibrating like a bell from the abrupt release of tidal stretching. "I did it," I shout. "It worked! Goddamn it, it really worked!"

This was what was predicted by the theorists, that I would be able to pass through the wormhole before it was shredded by the singularity at the center. The other possibility, that the singularity itself, infinitesimally small and infinitely powerful, might follow me through the wormhole, was laughed at by everyone who had any claim to understand wormhole physics. This time, the theorists were right.

But where am I?

There should be congratulations pouring into my radio by now, teams of friends and technicians swarming over to greet me, cheering and shouting.

"*Huis Clos,*" I say, over the radio. "I made it! *Huis Clos* here. Is anybody there?"

In theory, I should have reemerged at Wolf-562. But I do not see it. In fact, what I see is not recognizably the universe at all.

There are no stars.

Instead of stars, the sky is filled with lines, parallel lines of white light by the uncountable thousands. Dominating the sky, where the star Wolf-562 should have been, is a glowing red cylinder, perfectly straight, stretching to infinity in both directions.

Have I been transported into some other universe? Could the black hole's gravity sever the wormhole, cutting it loose from our universe entirely, and connect it into this strange new one?

If so, it has doomed me. The wormhole behind me, my only exit from this strange universe, is already destroyed. Not that escaping through it could have done me any good—it would only have brought me back to the place I escaped, to be crushed by the singularity of the black hole.

I could just turn my brain off, and I will have lost nothing, in a sense. They will bring you out of your suspended state, tell you that the edition of you that dropped into the black hole failed to upload, and they lost contact after it passed the event horizon. The experiment failed, but you had never been in danger.

But, however much you think we are the same, *I am not you.* I am a unique individual. When they revive you, without your expected new memories, I will still be gone.

I want to survive, I want to return.

A universe of tubes of light! Brilliant bars of an infinite cage. The bright lines in the sky have slight variations in color, from pale red to plasma-arc blue. They must be similar to the red cylinder near me, I figure, but light-years away. How could a universe have lines of light instead of stars?

I am amazingly well equipped to investigate that question, with senses that range from radio through X ray, and I have nothing else to do for the next thousand years or so. So I take a spectrum of the light from the glowing red cylinder.

I have no expectation that the spectrum will reveal anything I can interpret, but oddly, it looks normal. Impossibly, it looks like the spectrum of a star.

The computer can even identify, from its data of millions of spectra, precisely which star. The light from the cylinder has the spectral signature of Wolf-562.

Coincidence? It cannot possibly be coincidence, out of billions of possible spectra, that this glowing sword in the sky has exactly the spectrum of the star that should have been there. There can be no other conclusion but that the cylinder *is* Wolf-562.

I take a few more spectra, this time picking at random three of the lines of light in the sky, and the computer analyzes them for me. A bright one: the spectrum of 61 Virginis. A dimmer one: a match to Wolf-1061. A blue-white streak: Vega.

The lines in the sky are stars.

What does this mean?

I'm not in another universe. I am in *our* universe, but the universe has been transformed. Could the collision of a wormhole with a black hole destroy our entire universe, stretching suns like taffy into infinite straight lines? Impossible. Even if it had, I would still see faraway stars as dots, since the light from them has been traveling for hundreds of years.

The universe cannot have changed. Therefore, by logic, it must be *me* who has been transformed.

Having figured out this much, the only possible answer is obvious.

When the mathematicians describe the passage across the event horizon of a black hole, they say that the space and time directions switch identity. I had always thought this only a mathematical oddity, but if it were true, if I had rotated when I passed the event horizon, and was now perceiving time as a direction in space, and one of the space axes as time—this would explain everything. Stars extend from billions of years into the past to long into the future; perceiving time as space, I see lines of light. If I were to come closer and find one of the rocky planets of Wolf-562, it would look like a braid around the star, a helix of solid rock. Could I land on it? How would I interact with a world where what I perceive as time is a direction in space?

My physicist persona doesn't like this explanation, but is at a loss to find a better one. In this strange sideways existence, I must be violating the conservation laws of physics like mad, but the persona could find no other hypothesis and must reluctantly agree: Time is rotated into space.

To anybody outside, I must look like a string, a knobby long rope with one end at the wormhole and the other at my death, wherever that might be. But nobody could see me fast enough, since with no extension in time I must only be a transient even that bursts everywhere into existence and vanishes at the same instant. There is no way I can signal, no way I can communicate—

Or? Time, to me, is now a direction I can travel in as simply as using my rocket. I could find a planet, travel parallel to the direction of the surface—

But, no, all I could do would be to appear to the inhabitants briefly as a disk, a cross section of myself, infinitely thin. There is no way I could communicate.

But I can travel in time, if I want. Is there any way I can use this?

Wait. If I have rotated from space into time, then there is one direction in space that I cannot travel. Which direction is that? The direction that used to be away from the black hole.

Interesting thoughts, but not ones which help me much. To return, I need to once again flip space and time. I could dive into a black hole. This would again rotate space and time, but it wouldn't do me any good: Once I left the black hole—if I could leave the black hole—nothing would change.

Unless there were a wormhole inside the black hole, falling inward to destruction just at the same instant I was there? But the only wormhole that has fallen into a black hole was already destroyed. Unless, could I travel forward in time? Surely someday the research team would drop a new wormhole into the black hole—

Idiot. Of course there's a solution. Time is a spacelike dimension to me, so I can travel either direction in time now, forward or back. I need only to move back to an instant just after the wormhole passed through the event horizon, and, applying full thrust, shoot through. The very moment that my original self shoots through the wormhole to escape the singularity, I can pass through the opposite direction, and rotate myself back into the real universe.

The station at Virgo black hole is forty light-years away, and I don't dare use the original wormhole to reach it. My spacetime-rotated body must be an elongated snake in this version of space-time, and I do not wish to find out what a wormhole passage will do to it until I have no other choice. Still, that is no problem for me. Even with barely enough fuel to thrust for a few microseconds, I can reach an appreciable fraction of light-speed, and I can slow down my brain to make the trip appear only an instant.

To an outside observer, it takes literally no time at all.

• • •

"No," says the psych tech, when I ask her. "There's no law that compels you to uplink back into your original. You're a free human being. Your original can't force you."

"Great," I say. Soon I'm going to have to arrange to get a biological body built for myself. This one is superb, but it's a disadvantage in social intercourse being only a millimeter tall.

The transition back to real space worked perfectly. Once I figured out how to navigate in time-rotated space, it had been easy enough to find the wormhole and the exact instant it had penetrated the event horizon.

"Are you going to link your experiences to public domain?" the tech asks. "I think he would like to see what you experienced. Musta been pretty incredible."

"Maybe," I said.

"For that matter," the psych tech added, "I'd like to link it, too."

"I'll think about it."

So I am a real human being now, independent of you, my original.

There had been cheers and celebrations when I had emerged from the wormhole, but nobody had an inkling quite how strange my trip had been until I told them. Even then, I doubt that I was quite believed until the sensor readings and computer logs of *Huis Clos* confirmed my story with hard data.

The physicists had been ecstatic. A new tool to probe time and space. The ability to rotate space into time will open up incredible capabilities. They were already planning new expeditions, not the least of which was a trip to probe right to the singularity itself.

They had been duly impressed with my solution to the problem, although, after an hour of thinking it over, they all agreed it had been quite obvious. "It was lucky," one of them remarked, "that you decided to go through the wormhole from the opposite side, that second time."

"Why?" I asked.

"If you'd gone through the same direction, you'd have rotated an additional ninety degrees, instead of going back."

"So?"

"Reversed the time vector. Turns you into antimatter. First touch of the interstellar medium—Poof."

"Oh," I said. I hadn't thought of that. It made me feel a little less clever.

Now that the mission is over, I have no purpose, no direction for my existence. The future is empty, the black hole that we all must travel into. I will get a biological body, yes, and embark on the process of finding out who I am. Maybe, I think, this is a task that everybody has to do.

And then I will meet you. With luck, perhaps I'll even like you.

And maybe, if I should like you enough, and I feel confident, I'll decide to upload you into myself, and once more, we will again be one.

THE GRAVITY MINE

Stephen Baxter

Like many of his colleagues here at the beginning of a new century, British writer Stephen Baxter has been engaged for the last ten years or so with the task of revitalizing and reinventing the "hard-science" story for a new generation of readers, producing work on the Cutting Edge of science that bristles with weird new ideas and often takes place against vistas of almost outrageously cosmic scope.

Baxter made his first sale to Interzone *in 1987, and since then has become one of that magazine's most frequent contributors, as well as making sales to* Asimov's Science Fiction, Science Fiction Age, Zenith, New Worlds, *and elsewhere. He's one of the most prolific new writers in science fiction, and is rapidly becoming one of the most popular and acclaimed of them as well. In 2001, he appeared on the Final Hugo Ballot twice (once with the story that follows), and won both the* Asimov's *Readers Award and* Analog's *Analytical Laboratory Award, one of the few writers ever to win both awards in the same year. Baxter's first novel,* Raft, *was released in 1991 to wide and enthusiastic response, and was rapidly followed by other well-received novels such as* Timelike Infinity, Anti-Ice, Flux, *and the H. G. Wells pastiche—a sequel to* The Time Machine—The Time Ships, *which won both the John W. Campbell Memorial Award and the Philip K. Dick Award. His other books include the novels,* Voyage, Titan, Moonseed, *and* Mammoth, Book One: Silverhair, *and the collections* Vacuum Diagrams: Stories of the Xeelee Sequence, *and* Traces. *His most recent books are the novels* Manifold: Time, Manifold: Space, *and a novel written in*

collaboration with Arthur C. Clark, The Light of Other
Days.

*Here he takes us deeper into the future and farther
away from the flesh and the limitations and constraints of
human form than anyone else, even in this farsighted anthology, sweeping us along on a cosmic journey of mind-
blowing scope an scale, introducing us to characters who
no longer even remember what it was like to live as mortal
flesh and blood—or, indeed, even that they ever* did. *Which
isn't to say that they don't still have* problems . . .

*C*all her Anlic.

The first time she woke, she was in the ruins of an abandoned gravity mine.

At first the Community had chased around the outer
strata of the great gloomy structure. But at last, close to the
core, they reached a cramped ring. Here the central black
hole's gravity was so strong that light itself curved in
closed orbits.

The torus tunnel looked infinitely long. And they could
race as fast as they dared.

As they hurtled past fullerene walls, they could see
multiple images of themselves, a glowing golden mesh before and behind, for the echoes of their light endlessly circled the central knot of spacetime. "Just like the old days!"
they called, excited. "Just like the Afterglow . . . !"

Exhilarated, they pushed against the light barrier, and
those trapped circling images shifted to blue or red.

That was when it happened.

This Community was just a small tributary of the Conflux: isolated here in this ancient place, the density of mind
already stretched thin. And now, as light-speed neared, that
isolation stretched to breaking point.

. . . She budded off from the rest, her consciousness

made discrete, separated from the greater flow of minds and memories.

She slowed. The others rushed on without her, a dazzling circular storm orbiting the exhausted black hole. It felt like coming awake, emerging from a dream.

Her questions were immediate, flooding her raw mind. "Who am I? How did I get here?" And so on. The questions were simple, even trite. And yet they were unanswerable.

Others gathered around her—curious, sympathetic— and the race of streaking light began to lose its coherence.

One of them came to her.

Names meant little; this "one" was merely a transient sharpening of identity from the greater distributed entity that made up the Community.

Still, here he was. Call him Gaedor.

". . . Anlic?"

"I feel—odd," she said.

"Don't worry."

"Who am I?"

"Come back to us."

He reached for her, and she sensed the warm depths of companionship and memory and shared joy that lay beyond him. Depths waiting to swallow her up, to obliterate her questions.

She snapped, "No!" And, willfully, she sailed up and out and away, passing through the thin walls of the tunnel.

At first it was difficult to climb out of this twisted gravity well. But soon she was rising through layers of structure.

Here was the tight electromagnetic cage that had once tapped the spinning black hole like a dynamo. Here was the cloud of compact masses that had been hurled along complex orbits through the hole's ergosphere, extracting gravitational energy. It was antique engineering, long abandoned.

She emerged into a blank sky, a sky stretched thin by the endless expansion of spacetime.

Geador was here. "What do you see?"

"Nothing."

"Look harder." He showed her how.

There was a scattering of dull red pinpoints all around the sky.

"They are the remnants of stars," he said.

He told her about the Afterglow: that brief, brilliant period after the Big Bang, when matter gathered briefly in clumps and burned by fusion light. "It was a bonfire, over almost as soon as it began. The universe was very young. It has swollen some ten thousand trillion times in size since then. . . . Nevertheless, it was in that gaudy era that humans arose. *Us,* Anlic."

She looked into her soul, seeking warm memories of the Afterglow. She found nothing.

She looked back at the gravity mine.

At its center was a point of yellow-white light. Spears of light arced out from its poles, knife-thin. The spark was surrounded by a flattened cloud, dull red, inhomogeneous, clumpy. The big central light cast shadows through the crowded space around it.

It was beautiful, a sculpture of light and crimson smoke.

"This is Mine One," Geador said gently. "The first mine of all. And it is built on the ruins of the primeval galaxy—the galaxy from which humans first emerged."

"The first galaxy?"

"But it was all long ago." He moved closer to her. "So long ago that this mine became exhausted. Soon it will evaporate away completely. We have long since had to move on . . ."

But that had happened before. After all, humans had started from a single star, and spilled over half the universe, even before the stars ceased to shine.

Now humans wielded energy, drawn from the great

gravity mines, on a scale unimagined by their ancestors. Of course mines would be exhausted—like this one—but there would be other mines. Even when the last mine began to fail, they would think of *something*.

The future stretched ahead, long, glorious. Minds flowed together in great rivers of consciousness. There was immortality to be had, of a sort, a continuity of identity through replication and confluence across trillions upon trillions of years.

It was the Conflux.

Its source was far upstream.

The crudities of birth and death had been abandoned even before the Afterglow was over, when man's biological origins were decisively shed. So every mind, every tributary that made up the Conflux today had its source in that bright, remote upstream time.

Nobody had been born since the Afterglow.

Nobody but Anlic

". . . Come back," Geador said.

Her defiance was dissipating.

She understood nothing about herself. But she didn't want to be different. She didn't want to be unhappy.

There wasn't anybody who was less than maximally happy, the whole of the time. Wasn't that the purpose of existence?

So, troubled, she gave herself up to Geador, to the Conflux. And, along with her identity, her doubts and questions dissolved.

The universe would grow far older before she woke again.

". . . Flee! Faster! As fast as you can . . . !"

There was turbulence in the great rushing river of mind.

And in that turbulence, here and there, souls emerged

from the background wash. Each brief fleck suffered a moment of terror before falling back into the greater dreaming whole.

One of those flecks was Anlic.

In the sudden dark she clung to herself. She slithered to
a stop.

Transient identities clustered around her. "What are you
doing? Why are you staying here? You will be harmed."
They sought to absorb her, but fell back, baffled by her resistance.

The Community was fleeing, in panic. Why?

She looked back.

There was something there, in the greater darkness. She
made out the faintest of patterns: charcoal gray on black,
almost beyond her ability to resolve it, a mesh of neat regular triangles covering the sky. Visible through the inter-
stices was a complex, textured curtain of gray-pink light.

It was a structure that spanned the universe.

She felt stunned, disoriented. It was so different from
Mine One, her last clear memory. She must have crossed a
great desert of time.

But—she found, when she looked into her soul—her
questions remained unanswered.

She called out: "Geador?"

A ripple of shock and doubt spread through the Community.

". . . Are you Anlic."

"Geador?"

"I have Geador's memories."

That would have to do, she thought, irritated; in the
Conflux, memory and identity were fluid, distributed, am-
biguous.

"We are in danger, Anlic. You must come."

She refused to comply, stubborn. She indicated the
great netting. "Is that Mine One?"

"No" he said sadly. "Mine One was long ago, child."

"*How* long ago?"

"Time is nested . . ."

From this vantage, the era of man's first black hole empire had been the springtime, impossibly remote. And the Afterglow itself—the star-burning dawn—was lost, a mere detail of the Big Bang.

"What is happening here, Geador?"

"There is no time—"

"Tell me."

The universe had ballooned, fueled by time, and its physical processes had proceeded relentlessly.

Just as each galaxy's stars had dissipated, leaving a rump that had collapsed into a central black hole, so clusters of galaxies had broken up, and the remnants fell inward to cluster-scale holes. And the clusters in turn collapsed into supercluster-scale holes—the largest black holes to have formed naturally, with masses of a hundred trillion stars.

These were the cold hearths around which mankind now huddled.

"But," said Geador, "the supercluster holes are evaporating away—dissipating in a quantum whisper, like all black holes. The smallest holes, of stellar mass, vanished when the universe was a fraction of its present age. Now the largest natural holes, of supercluster mass, are close to exhaustion as well. And so we must farm them.

"Look at the City." He meant the universe-spanning net, the rippling surfaces within.

The City was a netted sphere. It contained giant black holes, galactic supercluster mass and above. They had been deliberately assembled. And they were merging, in a hierarchy of more and more massive holes. Life could subsist on the struts of the City, feeding off the last trickle of free energy.

Mankind was *moving* supercluster black holes, coalescing them in hierarchies all over the reachable universe, seeking to extend their lifetimes. It was a great challenge.

Too great.

Somberly, Geador showed her more.

The network was disrupted. It looked as if some immense object had punched out from the inside, ripping and twisting the struts. The tips of the broken struts were glowing a little brighter than the rest of the network, as if burning. Beyond the damaged network she could see the giant coalescing holes, their horizons distorted, great frozen waves of infalling matter visible in their cold surfaces.

This was an age of war: an obliteration of trillion-year memories, a bonfire of identity. Great rivers of mind were guttering, drying.

"This is the Conflux. How can there be war?"

Geador said, "We are managing the last energy sources of all. We have responsibility for the whole of the future. With such responsibility comes tension, disagreement. Conflict." She sensed his gentle, bitter humor. "We have come far since the Afterglow, Anlic. But in some ways we have much in common with the brawling argumentative apes of that brief time."

"Apes . . . ? Why am I here, Geador?"

"You're an eddy in the Conflux. We all wake up from time to time. It's just an accident. Don't trouble, Anlic. You are not alone. You have us."

Deliberately she moved away from him. "But I am not like you," she said bleakly. "*I* do not recall the Afterglow. I don't know where I came from."

"What does it matter?" he said harshly. "You have existed for all but the briefest moments of the universe's long history—"

"Has there been another like me?"

He hesitated. "No," he said. "No other like you. There hasn't been long enough."

"Then I *am* alone."

"Anlic, all your questions will be over, answered or not, if you let yourself die here. Come now . . ."

She knew he was right.

She fled with him. The great black hole City disappeared behind her, its feeble glow attenuated by her gathering velocity.

She yielded to Geador's will. She had no choice. Her questions were immediately lost in the clamor of community.

She would wake only once more.

Start with a second.

Zoom out. Factor it up to get the life of the Earth, with that second a glowing moment embedded within. Zoom out *again,* to get a new period, so long Earth's lifetime is reduced to the span of that second. Then nest it. Do it again. And again and again and *again* . . .

Anlic, for the last time, came to self-awareness.

It was inevitable that, given enough time, she would be budded by chance occurrence. And so it happened.

She clung to herself and looked around.

It was dark here. Vast, wispy entities cruised across spacetime's swelling breast.

There were no dead stars, no rogue planets. The last solid matter had long evaporated: burned up by proton decay, a thin smoke of neutrinos drifting out at lightspeed.

For ages the black hole engineers had struggled to maintain their Cities, to gather more material to replace what decayed away. It was magnificent, futile.

The last structures failed, the last black holes allowed to evaporate.

The Conflux of minds had dispersed, flowing out over the expanding universe like water running into sand.

Even now, of course, there was *something* rather than nothing. Around her was an unimaginably thin plasma:

free electrons and positrons decayed from the last of the Big Bang's hydrogen, orbiting in giant, slow circles. This cold soup was the last refuge of humanity.

The others drifted past her like clouds, immense, slow, coded in wispy light-year-wide atoms. And even now, the others clung to the solace of community.

But that was not for Anlic.

She pondered for a long time, determined not to slide back into the eternal dream.

At length she understood how she had come to be.

And she knew what she must do.

She sought out Mine One, the wreckage of man's original galaxy. The search took more empty ages.

With caution, she approached what remained.

There was no shape here. No form, no color, no time, no order. And yet there was motion: a slow, insidious, endless writhing, punctuated by bubbles that rose and burst, spitting out fragments of mass-energy.

This was the singularity that had once lurked within the great black hole's event horizon. Now it was naked, a glaring knot of quantum foam, a place where the unification of spacetime had been ripped apart to become a seething probabilistic froth.

Once this object had oscillated violently, and savage tides, chaotic and unpredictable, had torn at any traveler unwary enough to come close. But the singularity's energy had been dissipated by each such encounter.

Even singularities aged.

Still, the frustrated energy contained there seethed, quantum-mechanically, randomly. And sometimes, in those belched fragments, put there purely by chance, there were hints of order.

Structure. Complexity.

She settled herself around the singularity's cold glow.

Free energy was dwindling to zero, time stretching to

infinity. It took her longer to complete a single thought than it had once taken species to rise and fall on Earth.

It didn't matter. She had plenty of time.

She remembered her last conversation with Geador. *Has there been another like me? . . . No. No other like you. There hasn't been long enough.*

Now Anlic had all the time there was. The universe was exhausted of everything but time.

The longer she waited, the more complexity emerged from the singularity. Purely by chance. Much of it dissipated, purposeless.

But some of the mass-energy fragments had sufficient complexity to be able to gather and store information about the thinning universe. Enough to grow.

That, of course, was not enough. She continued to wait.

At last—by chance—the quantum tangle emitted a knot of structure sufficiently complex to reflect, not just the universe outside, but its own inner state.

Anlic moved closer, coldly excited.

It was a spark of consciousness: not descended from the grunting, breeding humans of the Afterglow, but born from the random quantum flexing of a singularity.

Just as she had been.

Anlic waited, nurturing, refining the rootless being's order and cohesion. And it gathered more data, developed sophistication.

At last it—*she*—could frame questions.

". . . Who am I? Who are you? Why are there two and not one?"

Anlic said, "I have much to tell you." And she gathered the spark in her attenuated soul.

Together, mother and daughter drifted away, and the river of time ran slowly into an ummarked sea.

REEF

Paul J. McAuley

Born in Oxford, England, in 1955, Paul J. McAuley now makes his home in London. A professional biologist for many years, he sold his first story in 1984, and has gone on to be a frequent contributor to Interzone, *as well as to markets such as* Amazing, The Magazine of Fantasy & Science Fiction, Asimov's Science Fiction, When the Music's Over, *and elsewhere.*

McAuley is considered to be one of the best of the new breed of British writers (although a few Australian writers could be fit in under this heading as well) who are producing that brand of rigorous hard science fiction with updated modern and stylistic sensibilities that is sometimes referred to as "radical hard science fiction," but he also writes Dystopian sociological speculations about the very near future, and he also is one of the major young writers who are producing that revamped and retooled widescreen Space Opera that has sometimes been called the New Baroque Space Opera, reminiscent of the Superscience stories of the '30s taken to an even higher level of intensity and scale. His first novel, Four Hundred Billion Stars, *won the Philip K. Dick Award, and his acclaimed novel* Fairyland *won both the Arthur C. Clarke Award and the John W. Campbell Award in 1996. His other books include the novels* Of The Fall, Eternal Light, *and* Pasquale's Angel, *Confluence—a major trilogy of ambitious scope and scale set ten million years in the future, comprised of the novels* Child of the River, Ancient of Days, *and* Shrine of Stars— *and* Life on Mars. *His short fiction has been collected in* The King of the Hill and Other Stories *and* The Invisible Country, *and he is the co-editor, with Kim Newman, of an original anthology,* In Dreams. *His most recent book is a new novel,* Whole Wide World.

In the suspenseful and inventive story that follows, he suggests that even those who explore only by proxy may still feel a proprietary sense of obligation about the discoveries they make—and that even explorers in robot extensor bodies may still run into danger, conflict, and life-or-death adventure.

Margaret Henderson Wu was riding a proxy by telepresence deep inside Tigris Rift when Dzu Sho summoned her. The others in her crew had given up one by one and only she was left, descending slowly between rosy, smoothly rippled cliffs scarcely a hundred meters apart. These were pavements of the commonest vacuum organism, mosaics made of hundreds of different strains of the same species. Here and there bright red whips stuck out from the pavement; a commensal species that deposited iron sulphate crystals within its integument. The pavement seemed to stretch endlessly below her. No probe or proxy had yet reached the bottom of Tigris Rift, still more than thirty kilometers away. Microscopic flecks of sulfur-iron complexes, sloughed cells, and excreted globules of carbon compounds and other volatiles formed a kind of smog or snow, and the vacuum organisms deposited nodes and intricate lattices of reduced metals that, by some trick of superconductivity, produced a broad band electromagnetic resonance that pulsed like a giant's slow heartbeat.

All this futzed the telepresence link between operators and their proxies. One moment Margaret was experiencing the three-hundred-twenty-degree panorama of the little proxy's microwave radar, the perpetual tug of vacuum on its mantle, the tang of extreme cold, a mere thirty degrees above absolute zero, the complex taste of the vacuum smog (burnt sugar, hot rubber, tar), the minute squirts of hydrogen from the folds of the proxy's puckered nozzle as it maintained its orientation relative to the cliff face during its descent, with its tentacles retracted in a tight ball around the relay piton. The next, she was back in her cradled body in warm blackness, phosphenes floating in her vision and white noise in her ears while the transmitter searched for a viable waveband, locked on and—*pow*—she was back, falling past rippled pink pavement.

The alarm went off, flashing an array of white stars over the panorama. Her number two, Srin Kerenyi, said in her ear, "You're wanted, boss."

Margaret killed the alarm and the audio feed. She was already a kilometer below the previous benchmark and she wanted to get as deep as possible before she implanted the telemetry relay. She swiveled the proxy on its long axis, increased the amplitude of the microwave radar. Far below were intimations of swells and bumps jutting from the plane of the cliff face, textured mounds like brain coral, randomly orientated chimneys. And something else, clouds of organic matter perhaps—

The alarm again. Srin had overridden the cut-out.

Margaret swore and dove at the cliff, unfurling the proxy's tentacles and jamming the piton into pinkness rough with black papillae, like a giant's tongue quick frozen against the ice. The piton's spikes fired automatically. Recoil sent the little proxy tumbling over its long axis until it reflexively stabilized itself with judicious squirts of gas. The link rastered, came back, cut out completely. Margaret hit the switch that turned the tank into a chair; the mask lifted away from her face.

Srin Kerenyi was standing in front of her. "Dzu Sho wants to talk with you, boss. Right now."

•   •   •

The job had been offered as a sealed contract. Science crews had been informed of the precise nature of their tasks only when the habitat was under way. But it was good basic pay with the promise of fat bonuses on completion: When she had won the survey contract Margaret Henderson Wu had brought with her most of the crew from her previous job, and had nursed a small hope that this would be a change in her family's luck.

The *Ganapati* was a new habitat founded by an alliance of two of the Commonwealth's oldest patrician families. It was of standard construction, a basaltic asteroid cored by a gigawatt X-ray laser and spun up by vented rock vapor to give 0.2 gee on the inner surface of its hollowed interior, factories and big reaction motors dug into the stern. With its AIs rented out for information crunching and its refineries synthesizing exotic plastics from cane sugar biomass and gengeneered oilseed rape precursors, the new habitat had enough income to maintain the interest on its construction loan from the Commonwealth Bourse, but not enough to attract new citizens and workers. It was still not completely fitted out, had less than a third of its optimal population.

Its Star Chamber, young and cocky and eager to win independence from their families, had taken a big gamble. They were chasing a legend.

Eighty years ago, an experiment in accelerated evolution of chemoautotrophic vacuum organisms had been set up on a planetoid in the outer edge of the Kuiper Belt. The experiment had been run by a shell company registered on Ganymede but covertly owned by the Democratic Union of China. In those days, companies and governments of Earth had not been allowed to operate in the Kuiper Belt, which had been claimed and ferociously defended by outer system cartels. That hegemony had ended in the Quiet War, but the Quiet War had also destroyed all records of

the experiment; even the Democratic Union of China had
disappeared, absorbed into the Pacific Community.

There were over fifty thousand objects with diameters
greater than a hundred kilometers in the Kuiper Belt, and a
billion more much smaller, the plane of their orbits stretching beyond those of Neptune and Pluto. The experimental
planetoid, Enki, named for one of the Babylonian gods of
creation, had been lost among them. It had become a legend, like the Children's Habitat, or the ghost comet, or the
pirate ship crewed by the reanimated dead, or the worker's
paradise of Fiddler's Green.

And then, forty-five years after the end of the Quiet
War, a data miner recovered enough information to recon-
struct Enki's eccentric orbit. She sold it to the *Ganapati*.
The habitat bought time on the Uranus deep space tele-
scopic array and confirmed that the planetoid was where it
was supposed to be, currently more than seven thousand
million kilometers from the Sun.

Nothing more was known. The experiment might have
failed almost as soon as it begun, but potentially it might
win the *Ganapati* platinum-rated credit on the Bourse.
Margaret and the rest of the science crews would, of
course, receive only their fees and bonuses, less deduc-
tions for air and food and water taxes, and anything they
bought with scrip in the habitat's stores; the indentured
workers would not even get that. Like every habitat in the
Commonwealth, the *Ganapati* was structured like an ancient Greek Republic, ruled by share-holding citizens who
lived in the landscaped parklands of the inner surface, and
run by indentured and contract workers who were housed
in the undercroft of malls and barracks tunneled into the
Ganapati's rocky skin.

On the long voyage out, the science crews had been on
minimal pay, far lower than that of the unskilled techs who
worked the farms and refineries, and the servants who
maintained the citizens' households. There were food

shortages because so much biomass was being used to make exportable biochemicals; any foodstuffs other than basic rations were expensive, and prices were carefully manipulated by the habitat's Star Chamber. When the *Ganapati* reached Enki and the contracts of the science crews were activated, food prices had increased accordingly. Techs and household servants suddenly found themselves unable to afford anything other than dole yeast. Resentment bubbled over into skirmishes and knife-fights, and a small riot the White Mice, the undercroft's police, subdued with gas. Margaret had to take time off to bail out several of her crew, had given them an angry lecture about threatening everyone's bonuses.

"We got to defend our honor," one of the men said.

"Don't be a fool," Margaret told him. "The citizens play workers against science crews to keep both sides in their places, and still turn a good profit from increases in food prices. Just be glad you can afford the good stuff now, and keep out of trouble."

"They were calling you names, boss," the man said. "On account you're—"

Margaret stared him down. She was standing on a chair, but even so she was a good head shorter than the gangling outers. She said, "I'll fight my own fights. I always have. Just think of your bonuses and keep quiet. It will be worth it. I promise you."

And it was worth it, because of the discovery of the reef.

At some time in the deep past, Enki had suffered an impact that had remelted it and split it into two big pieces and thousands of fragments. One lone fragment still orbited Enki, a tiny moonlet where the AI that had controlled the experiment had been installed; the others had been drawn together again by their feeble gravity fields, but had cooled before coalescence had been completed, leaving a vast deep chasm, Tigris Rift, at the lumpy equator.

Margaret's crew had discovered that the vacuum organisms had proliferated wildly in the deepest part of the Rift, deriving energy by oxidation of elemental sulfur and ferrous iron, converting, carbonaceous material into useful organic chemicals. There were crusts and sheets, things like thin scarves folded into fragile vases and chimneys, organ pipe clusters, whips, delicate fretted laces. Some fed on others, one crust slowly overgrowing and devouring another. Others appeared to be parasites, sending complex veins ramifying through the thalli of their victims. Water-mining organisms recruited sulfur oxidizers, trading precious water for energy and forming warty outgrowths like stromatolites. Some were more than a hundred meters across, surely the largest prokaryotic colonies in the known Solar System.

All this variety, and after only eighty years of accelerated evolution! Wild beauty won from the cold and the dark. The potential to feed billions. The science crews would get their bonuses, all right; the citizens would become billionaires.

Margaret spent all her spare time investigating the reef by proxy, pushed her crew hard to overcome the problems of penetrating the depths of the Rift. Although she would not admit it even to herself, she had fallen in love with the reef. She would gladly have explored it in person, but as in most habitats, the *Ganapati*'s citizens did not like their workers going where they themselves would not.

Clearly, the experiment had far exceeded its parameters, but no one knew why. The Ai that had overseen the experiment had shut down thirty years ago. There was still heat in its crude proton beam fission pile, but it had been overgrown by the very organisms it had manipulated.

Its task had been simple. Colonies of a dozen species of slow-growing chemo-autotrophs had been introduced into a part of the Rift rich with sulfur and ferrous iron. Thousands of random mutations had been induced. Most colonies had died, and those few which had thrived had been sampled, mutated, and reintroduced in a cycle repeated every hundred days.

But the AI had selected only for fast growth, not for adaptive radiation, and the science crews held heated seminars about the possible cause of the unexpected richness of the reef's biota. Very few believed that it was simply a result of accelerated evolution. Many terrestrial bacteria divided every twenty minutes in favorable conditions, and certain species were known to have evolved from being resistant to an antibiotic to becoming obligately dependent upon it as a food source in less than five days, or only three hundred and sixty generations, but that was merely a biochemical adaptation. The fastest division rate of the vacuum organisms in the Rift was less than once a day, and while that still meant more than thirty thousand generations had passed since the reef had been seeded, half a million years in human terms, the evolutionary radiation in the reef was the equivalent of Neanderthal Man evolving to fill every mammalian niche from bats to whales.

Margaret's survey crew had explored and sampled the reef for more than thirty days. Cluster analysis suggested that they had identified less than ten percent of the species that had formed from the original seed population. And now deep radar suggested that there were changes in the unexplored regions in the deepest part of Tigris Rift, which the proxies had not yet been able to reach.

Margaret had pointed this out at the last seminar. "We're making hypotheses on incomplete information. We don't know everything that's out there. Sampling suggests that complexity increases away from the surface. There could be thousands more species in the deep part of the Rift."

At the back of the room, Opie Kindred, the head of the genetics crew, said languidly, "We don't need to know everything. That's not what we're paid for. We've already

found several species that perform better than present
commercial cultures. The *Ganapati* can make money from
them and we'll get full bonuses. Who cares how they got
there?"

Arn Nivedta, the chief of the biochemist crew, said,
"We're all scientists here. We prove our worth by finding
out how things work. Are your mysterious experiments no
more than growth tests, Opie? If so, I'm disappointed."

The genetics crew had set up an experimental station on
the surface of the *Ganapati,* off limits to everyone else.

Opie smiled. "I'm not answerable to you."

This was greeted with shouts and jeers. The science
crews were tired and on edge, and the room was hot and
poorly ventilated.

"Information should be free," Margaret said. "We all
work toward the same end. Or are you hoping for extra
bonuses, Opie?"

There was a murmur in the room. It was a tradition
that all bonuses were pooled and shared out between the
various science crews at the end of a mission.

Opie Kindred was a clever, successful man, yet somehow soured, as if the world was a continual disappoint-
ment. He rode his team hard, was quick to find failure in
others. Margaret was a natural target for his scorn, a squat
muscle-bound unedited dwarf from Earth who had to take
drugs to survive in micro-gravity, who grew hair in all
sorts of unlikely places. He stared at her with disdain and
said, "I'm surprised at the tone of this briefing, Dr. Wu.
Wild speculations built on nothing at all. I have sat here for
an hour and heard nothing useful. We are paid to get results, not generate hypotheses. All we hear from your crew
is excuses when what we want are samples. It seems simple enough to me. If something is upsetting your proxies,
then you should use robots. Or send people in and hand-
pick samples. I've worked my way through almost all

you've obtained. I need more material, especially in light of my latest findings."

"Robots need transmission relays, too," Srin Kerenyi pointed out.

Orly Higgins said, "If you ride them, to be sure. But I don't see the need for human control. It is a simple enough task to program them to go down, pick up samples, return." She was the leader of the crew that had unpicked the AI's corrupted code, and was an acolyte of Opie Kindred.

"The proxies failed whether or not they were remotely controlled," Margaret said, "and on their own they are as smart as any robot. I'd love to go down there myself, but the Star Chamber has forbidden it for the usual reasons. They're scared we'll get up to something if we go where they can't watch us."

"Careful, boss," Srin Kerenyi whispered. "The White Mice are bound to be monitoring this."

"I don't care," Margaret said. "I'm through with trying polite requests. We need to get down there, Srin."

"Sure, boss. But getting arrested for sedition isn't the way."

"There's some interesting stuff in the upper levels," Arn Nivedta said. "Stuff with huge commercial potential, as you pointed out, Opie."

Murmurs of agreement throughout the crowded room. The reef could make the *Ganapati* the richest habitat in the Outer System, where expansion was limited by the availability of fixed carbon. Even a modest-sized comet nucleus, ten kilometers in diameter, say, and salted with only one hundredth of one percent carbonaceous material, contained fifty million tons of carbon, mostly as methane and carbon monoxide ice, with a surface dusting of tarry long-chain hydrocarbons. The problem was that most vacuum organisms converted simple carbon compounds into organic matter using the energy of sunlight captured by a variety of photosynthetic pigments, and so could only grow on the surfaces of planetoids. No one had yet developed vacuum organisms that, using other sources of energy, could efficiently mine planetoids interiors, but that was what accelerated evolution appeared to have produced in the reef. It could enable exploitation of the entire volume of objects in the Kuiper Belt, and beyond, in the distant Oort Cloud. It was a discovery of incalculable worth.

Arn Nivedta waited for silence, and added, "Of course, we can't know what the commercial potential is until the reef species have been fully tested. What about it, Opie?"

"We have our own ideas about commercial potential," Opie Kindred said. "I think you'll find that we hold the key to success here."

Boos and catcalls at this from both the biochemists and the survey crew. The room was polarizing. Margaret saw one of her crew unsheathe a sharpened screwdriver, and she caught the man's hand and squeezed it until he cried out. "Let it ride," she told him. "Remember that we're scientists."

"We hear of indications of more diversity in the depths, but we can't seem to get there. One might suspect," Opie said, his thin upper lip lifting in a supercilious curl, "sabotage."

"The proxies are working well in the upper part of the Rift," Margaret said, "and we are doing all we can to get them operative farther down."

"Let's hope so," Opie Kindred said. He stood, and around him his crew stood, too. "I'm going back to work, and so should all of you. Especially you, Dr. Wu. Perhaps you should be attending to your proxies instead of planning useless expeditions."

And so the seminar broke up in uproar, with nothing productive coming from it and lines of enmity drawn through the community of scientists.

"Opie is scheming to come out of this on top," Arn Nivedta said to Margaret afterward. He was a friendly, en-

thusiastic man, tall even for an outer, and as skinny as a
rail. He stopped in Margaret's presence, trying to reduce
the extraordinary difference between their heights. He
said, "He wants desperately to become a citizen, and so he
thinks like one."

"Well, my God, we all want to be citizens," Margaret
said. "Who wants to live like this?"

She gestured, meaning the crowded bar, its rock walls
and low ceiling, harsh lights and the stink of spilled beer
and too many people in close proximity. Her parents had
been citizens, once upon a time. Before their run of bad
luck. It was not that she wanted those palmy days back—
she could scarcely remember them—but she wanted more
than this.

She said, "The citizens sleep between silk sheets and
eat real meat and play their stupid games, and we have to
do their work on restricted budgets. The reef is the discovery of the century, Arn, but God forbid that the citizens
should begin to exert themselves. We do the work, they
fuck in rose petals and get the glory."

Arn laughed at this.

"Well, it's true!"

"It's true we have not been as successful as we might
like," Arn said mournfully.

Margaret said reflectively, "Opie's a bastard, but he's
smart, too. He picked just the right moment to point the
finger at me."

Loss of proxies was soaring exponentially, and the
proxy farms of the *Ganapati* were reaching a critical point.
Once losses exceeded reproduction, the scale of exploration would have to be drastically curtailed, or the seed
stock would have to be pressed into service, a gamble the
Ganapati could not afford to take.

And then, the day after the disastrous seminar, Margaret was pulled back from her latest survey to account

for herself in front of the chairman of the *Ganapati*'s Star
Chamber.

"We are not happy with the progress of your survey, Dr.
Wu," Dzu Sho said. "You promise much, but deliver little."

Margaret shot a glance at Opie Kindred, and he smiled
at her. He was immaculately dressed in a gold-trimmed
white tunic and white leggings. His scalp was oiled and his
manicured fingernails were painted with something that
split light into rainbows. Margaret, fresh from the tank,
wore loose, grubby work grays. There was sticky elec-
trolyte paste on her arms and legs and shaven scalp, the
reek of sour sweat under her breasts and in her armpits.

She contained her anger and said, "I have submitted
daily reports on the problems we encountered. Progress is
slow but sure. I have just established a relay point a full
kilometer below the previous datum point."

Dzu Sho waved this away. He lounged in a blue gel
chair, quite naked, as smoothly fat as a seal. He had a
round, hairless head and pinched features, like a
thumbprint on an egg. The habitat's lawyer sat behind him,
a young woman neat and anonymous in a gray tunic suit.
Margaret, Opie Kindred, and Arn Nivedta sat on low
stools, supplicants to Dzu Sho's authority. Behind them,
half a dozen servants stood at the edge of the grassy space.

This was in an arbor of figs, ivy, bamboo, and fast-
growing banyan at the edge of Sho's estate. Residential
parkland curved above, a patchwork of spindly, newly
planted woods and meadows and gardens. Flyers were out,
triangular rigs in primary colors pirouetting around the
weightless axis. Directly above, mammoths the size of
large dogs grazed an upside-down emerald-green field.
The parkland stretched away to the ring lake and its slosh

barrier, three kilometers in diameter, and the huge farms
that dominated the inner surface of the habitat. Fields of
lentils, wheat, cane fruits, tomatoes, rice, and exotic veg-
etables for the tables of the citizens, and fields and fields
and fields of sugarcane and oilseed rape for the biochemical industry and the yeast tanks.

Dzu Sho said, "Despite the poor progress of the survey
crew, we have what we need, thanks to the work of Dr.
Kindred. This is what we will discuss."

Margaret glanced at Arn, who shrugged. Opie Kindred's smile deepened. He said, "My crew has established
why there is so much diversity here. The vacuum organisms have invented sex."

"We know they have sex," Arn said. "How else could
they evolve?"

His own crew had shown that the vacuum organisms
could exchange genetic material through pili, microscopic
hollow tubes grown between cells or hyphal strands. It was
analogous to the way in which genes for antibiotic resistance spread through populations of terrestrial bacteria.

"I do not mean genetic exchange, but genetic recombi-
nation," Opie Kindred said. "I will explain."

The glade filled with flat plates of color as the geneticist
conjured charts and diagrams and pictures from his slate.
Despite her anger, Margaret quickly immersed herself in
the flows of data, racing ahead of Opie Kindred's clipped
explanations.

It was not normal sexual reproduction. There was no
differentiation into male or female, or even into comple-
mentary mating strains. Instead, it was mediated by a
species that aggressively colonized the thalli of others.
Margaret had already seen it many times, but until now she
had thought that it was merely a parasite. Instead, as Opie
Kindred put it, it was more like a vampire.

A shuffle of pictures, movies patched from hundreds of
hours of material collected by roving proxies. Here was a

colony of the black crustose species found all through the
explored regions of the Rift. Time speeded up. The crustose colony elongated its ragged perimeter in pulsing
spurts. As it grew, it exfoliated microscopic particles. Margaret's viewpoint spiraled into a close-up of one of the exfoliations, a few cells wrapped in nutrient storing strands.

Millions of these little packages floated through the
vacuum. If one landed on a host thallus, it injected its genetic payload into the host cells. The view dropped inside
one such cell. A complex of carbohydrate and protein
strands webbed the interior like intricately packed spider-
webs. Part of the striated cell wall drew apart and a packet
of DNA coated in hydrated globulins and enzymes burst
inward. The packet contained the genomes of both the parasite and its previous victim. It latched on to protein
strands and crept along on ratchetting microtubule claws
until it fused with the cell's own circlet of DNA.

The parasite possessed an enzyme that snipped strands
of genetic material at random lengths. These recombined,
forming chimeric cells that contained genetic information
from both sets of victims, with the predator species'
genome embedded among the native genes like an inter-
penetrating text.

The process repeated itself in flurries of coiling and un-
coiling DNA strands as the chimeric cells replicated. It was
a crude, random process. Most contained incomplete or
noncomplementary copies of the genomes and were unable to function, or contained so many copies that transcription was halting and imperfect. But a few out of every
thousand were viable, and a small percentage of those
were more vigorous than either of their parents. They grew
from a few cells to a patch, and finally overgrew the
parental matrix in which they were embedded. There were
pictures that showed every stage of this transformation in
a laboratory experiment.

"This is why I have not shared the information until

now," Opie Kindred said, as the pictures faded around him. "I had to ensure by experimental testing that my theory was correct. Because the procedure is so inefficient, we had to screen thousands of chimeras until we obtained a strain that overgrew its parent."

"A very odd and extreme form of reproduction," Arn said. "The parent dies so that the child might live."

Opie Kindred smiled. "It is more interesting than you might suppose."

The next sequence showed the same colony, now clearly infected by the parasitic species—leprous black spots mottled its pinkish surface. Again time speeded up. The spots grew larger, merged, shed a cloud of exfoliations.

"Once the chimera overgrows its parent," Opie Kindred said, "the genes of the parasite, which have been reproduced in every cell of the thallus, are activated. The host cells are transformed. It is rather like an RNA virus, except that the virus does not merely subvert the protein and RNA-making machinery of its host cell. It takes over the cell itself. Now the cycle is completed, and the parasite sheds exfoliations that will turn infect new hosts.

"Here is the motor of evolution. In some of the infected hosts, the parasitic genome is prevented from expression, and the host becomes resistant to infection. It is a variation of the Red Queen's race. There is an evolutionary pressure upon the parasite to evolve new infective forms, and then for the hosts to resist them, and so on. Meanwhile, the host species benefit from new genetic combinations that by selection incrementally improve growth. The process is random but continuous, and takes place on a vast scale. I estimate that millions of recombinant cells are produced each hour, although perhaps only one in ten million are viable, and of those only one in a million are significantly more efficient at growth than their parents. But this is more than sufficient to explain the diversity we have mapped in the reef."

Arn said, "How long have you known this, Opie?"

"I communicated my findings to the Star Chamber just this morning," Opie Kindred said. "The work has been very difficult. My crew has to work under very tight restraints, using Class Four containment techniques, as with the old immunodeficiency plagues."

"Yah, of course," Arn said. "We don't know how the exfoliations might contaminate the ship."

"Exactly" Opie Kindred said. "That is why the reef is dangerous."

Margaret bridled at this. She said sharply, "Have you tested how long the exfoliations survive?"

"There is a large amount of data about bacterial spore survival. Many survive thousands of years in vacuum close to absolute zero. It hardly seems necessary—"

"You didn't bother," Margaret said. "My God, you want to destroy the reef and you have no *evidence*. You didn't *think*."

It was the worst of insults in the scientific community. Opie Kindred colored, but before he could reply, Dzu Sho held up a hand, and his employees obediently fell silent.

"The Star Chamber has voted," Dzu Sho said. "It is clear that we have all we need. The reef is dangerous, and must be destroyed. Dr. Kindred has suggested a course of action that seems appropriate. We will poison the sulfur-oxidizing cycle and kill the reef."

"But we don't know—"

"We haven't found—"

Margaret and Arn had spoken at once. Both fell silent when Dzu Sho held up a hand again. He said, "We have isolated commercially useful strains. Obviously, we can't use the organisms we have isolated because they contain the parasite within every cell. But we can synthesize useful gene sequences and splice them into current commercial strains of vacuum organism to improve quality."

"I must object," Margaret said. "This is a unique construct. The chances of it evolving again are minimal. We must study it further. We might be able to discover a cure for the parasite."

"It is unlikely," Opie Kindred said. "There is no way to eliminate the parasite from the host cells by gene therapy because they are hidden within the host chromosome, shuffled in a different pattern in every cell of the trillions of cells that make up the reef. However, it is quite easy to produce a poison that will shut down the sulfur-oxidizing metabolism common to the different kinds of reef organism."

"Production has been authorized," Sho said. "It will take, what did you tell me, Dr. Kindred?"

"We require a large quantity, given the large biomass of the reef. Ten days at least. No more than fifteen."

"We have not studied it properly," Arn said. "So we cannot yet say what and what is not possible."

Margaret agreed, but before she could add her objection, her earpiece trilled, and Srin Kerenyi's voice said apologetically, "Trouble, boss. You better come at once."

The survey suite was in chaos, and there was worse chaos in the Rift. Margaret had to switch proxies three times before she found one she could operate. All around her, proxies were fluttering and jinking, as if caught in strong currents instead of floating in vacuum in virtual free fall.

This was at the four-thousand-meter level, where the nitrogen ice walls of the Rift were sparsely patched with yellow and pink marblings that followed veins of sulfur and organic contaminants. The taste of the vacuum smog here was strong, like burnt rubber coating Margaret's lips and tongue.

As she looked around, a proxy jetted toward her. It overshot and rebounded from a gable of frozen nitrogen, its nozzle jinking back and forth as it tried to stabilize its position.

"Fuck," its operator, Kim Nieye, said in Margaret's ear. "Sorry, boss. I've been through five of these, and now I'm losing this one."

On the other side of the cleft, a hundred meters away, two specks tumbled end for end, descending at a fair clip toward the depths. Margaret's vision color-reversed, went black, came back to normal. She said, "How many?"

"Just about all of them. We're using proxies that were up in the tablelands, but as soon as we bring them down, they start going screwy, too."

"Herd some up and get them to the sample pickup point. We'll need to do dissections."

"No problem, boss. Are you okay?"

Margaret's proxy had suddenly upended. She couldn't get its trim back. "I don't think so," she said, and then the proxy's nozzle flared and with a pulse of gas the proxy shot away into the depths.

It was a wild ride. The proxy expelled all its gas reserves, accelerating as straight as an arrow. Coralline formations blurred past, and then long stretches of sulfur-eating pavement. The proxy caromed off the narrowing walls and began to tumble madly.

Margaret had no control. She was a helpless but exhilarated passenger. She passed the place where she had set the relay and continued to fall. The link started to break up. She lost all sense of proprioception, although given the tumbling fall of the proxy, that was a blessing. Then the microwave radar started to go, with swathes of raster washing across the false color view. Somehow the proxy managed to stabilize itself, so it was falling headfirst toward the unknown regions at the bottom of the Rift. Margaret glimpsed structures swelling from the walls. And

then everything went away and she was back, sweating
and nauseous in the couch.

It was bad. More than ninety-five percent of the proxies
had been lost. Most, like Margaret's, had been lost in the
depths. A few, badly damaged by collision, had been
stranded among the reef colonies, but proxies sent to re-
trieve them went out of control, too. It was clear that some
kind of infective process had affected them. Margaret had
several dead proxies collected by a maintenance robot and
ordered that the survivors should be regrouped and kept
above the deep part of the Rift where the vacuum organisms proliferated. And then she went to her suite in the undercroft and waited for the Star Chamber to call her before
them.

The Star Chamber took away Margaret's contract, citing
failure to perform and possible sedition (that remark in the
seminar had been recorded). She was moved from her suite
to a utility room in the lower level of the undercroft and
put to work in the farms.

She thought of her parents.
She had been here before.
She thought of the reef.
She couldn't let it go.
She would save it if she could.

Srin Kerenyi kept her up to date. The survey crew and
its proxies were restricted to the upper level of the reef.
Manned teams under Opie Kindred's control were exploring the depths—*he* was trusted where Margaret was not—
but if they discovered anything it wasn't communicated to
the other science crews.

Margaret was working in the melon fields when Arn
Nivedta found her. The plants sprawled from hydroponic
tubes laid across gravel beds, beneath blazing lamps hung
in the axis of the farmlands. It was very hot, and there was

a stink of dilute sewage. Little yellow ants swarmed everywhere. Margaret had tucked the ends of her pants into the
rolled tops of her shoesocks, and wore a green eyeshade.
She was using a fine paintbrush to transfer pollen to the
stigma of the melon flowers.

Arn came bouncing along between the long rows of
plants like a pale scarecrow intent on escape. He wore only
light black shorts and a web belt hung with pens, little silvery tools and a notepad.

He said, "They must hate you, putting you in a shithole
like this."

"I have to work, Arn. Work or starve. I don't mind it. I
grew u working the fields."

Not strictly true: Her parents had been ecosystem de-
signers. But it was how it had ended.

Arn said cheerfully, "I'm here to rescue you. I can
prove it wasn't your fault."

Margaret straightened, one hand on the small of her
back where a permanent ache had lodged itself. She said,
"Of course it wasn't my fault. Are you all right?"

Arn had started to hop about, brushing at one bare long-
toed foot and then the other. The ants had found him. His
toes curled like fingers. The big toes were opposed. Mon-
key feet.

"Ants are having something of a population explosion,"
she said. "We're in the stage between introduction and sta-
bilization here. The cycles will smooth out as the ecosystem matures."

Arn brushed at his legs again. His prehensile big toe
flicked an ant from the sole of his foot. "They want to in-
corporate me into the cycle, I think."

"We're all in the cycle, Arn. The plants grow in sewage;
we eat the plants." Margaret saw her supervisor coming toward them through the next field. She said, "We can't talk
here. Meet me in my room after work."

•   •   •

Margaret's new room was barely big enough for a hammock, a locker, and a tiny shower with a toilet pedestal. Its rock walls were unevenly coated with dull green fiber spray. There was a constant noise of pedestrians beyond the oval hatch; the air-conditioning allowed in a smell of frying oil and ketones despite the filter trap Margaret had set up. She had stuck an aerial photograph of New York, where she had been born, above the head stay of her hammock, and dozens of glossy printouts of the reef scaled the walls. Apart from the pictures, a few clothes in the closet, and the spider plant under the purple grolite, the room was quite anonymous.

She had spent most of her life in rooms like this. She could pack in five minutes, ready to move onto the next job.

"This place is probably bugged," Arn said. He sat with his back to the door, sipping schnapps from a silvery flask and looking at the overlapping panoramas of the reef.

Margaret sat on the edge of her hammock. She was nervous and excited. She said, "Everywhere is bugged. I want them to hear that I'm not guilty. Tell me what you know."

Arn looked at her. "I examined the proxies you sent back. I wasn't quite sure what I was looking for, but it was surprisingly easy to spot."

"An infection," Margaret said.

"Yah, a very specific infection. We concentrated on the nervous system, given the etiology. In the brain we found lesions, always in the same area."

Margaret examined the three-dimensional color-enhanced tomographic scan Arn had brought. The lesions were little black bubbles in the underside of the unfolded cerebellum, just in front of the optic node.

"The same in all of them," Arn said. "We took samples, extracted DNA, and sequenced it." A grid of thousands of colored dots, then another superimposed over it. All the dots lined up.

"A match to Opie's parasite," Margaret guessed.

Arn grinned. He had a nice smile. It made him look like an enthusiastic boy. "We tried that first, of course. Got a match, then went through the library of reef organisms, and got partial matches. Opie's parasite has its fingerprints in the DNA of everything in the reef, but this"—he jabbed a long finger through the projection—"is the pure quill. Just an unlucky accident that it lodges in the brain at this particular place and produces the behavior you saw."

"Perhaps it isn't a random change," Margaret said. "Perhaps the reef has a use for the proxies."

"Teleology," Arn said. "Don't let Opie hear that thought. He'd use it against you. This is evolution. It isn't directed by anything other than natural selection. There is no designer, no watchmaker. Not after the AI crashed, anyway, and it only pushed the ecosystem toward more efficient sulfur oxidation. There's more, Margaret. I've been doing some experiments on the side. Exposing aluminum foil sheets in orbit around Enki. There are exfoliations everywhere."

"Then Opie is right."

"No, no. All the exfoliations I found were nonviable. I did more experiments. The exfoliations are metabolically active when released, unlike bacterial spores. And they have no protective wall. No reason for them to have one, yah? They live only for a few minutes. Either they land on a new host or they don't. Solar radiation easily tears them apart. You can kill them with a picowatt ultraviolet laser. Contamination isn't a problem."

"And it can't infect us," Margaret said. "Vacuum organisms and proxies have the same DNA code as us, the same as everything from Earth, for that matter, but it's written in artificial nucleotide bases. The reef isn't dangerous at all, Arn."

"Yah, but in theory it could infect every vacuum organism ever designed. The only way around it would be to

change the base structure of vacuum organism DNA—
how much would that cost?"

"I know about contamination, Arn. The mold that
wrecked the biome designed by my parents came in with
someone or something. Maybe on clothing, or skin, or in
the gut, or in some trade goods. It grew on anything with a
cellulose cell wall. Every plant was infected. The fields
were covered by huge sheets of gray mold; the air was full
of spores. It didn't infect people, but more than a hundred
died from massive allergic reactions and respiratory failure. They had to vent the atmosphere in the end. And my
parents couldn't find work after that."

Arn said gently, "That is the way. We live by our repu-
tations. It's hard when something goes wrong."

Margaret ignored this. She said, "The reef is a resource,
not a danger. You're looking at it the wrong way, like Opie
Kindred. We need diversity. Our biospheres have to be
complicated because simple systems are prone to invasion
and disruption, but they aren't one hundredth as complicated as those on Earth. If my parents' biome had been
more diverse, the mold couldn't have found a foothold."

"There are some things I could do without." Arn
scratched his left ankle with the toes of his right foot.
"Like those ants."

"Well, we don't know if we need the ants specifically,
but we need variety, and they contribute to it. They help
aerate the soil, to begin with, which encourages stratifica-
tion and diversity of soil organisms. There are a million
different kinds of microbe in a gram of soil from a forest
on Earth; we have to make do with less than a thousand.
We don't have one tenth that number of useful vacuum organisms and most are grown in monoculture, which is the
most vulnerable ecosystem of all. That was the cause of
the crash of the green revolution on Earth in the twenty-
first century. But there are hundreds of different species in
the reef. Wild species, Arn. You could seed a planetoid

with them and go harvest it a year later. The citizens don't go outside because they have their parklands, their palaces, their virtualities. They've forgotten that the outer system isn't just the habitats. There are millions of small planetoids in the Kuiper Belt. Anyone with a dome and the reef vacuum organisms could homestead one."

She had been thinking about this while working out in the fields. The Star Chamber had given her plenty of time to think.

Arn shook his head. "They all have the parasite lurking in them. Any species from the reef can turn into it. Perhaps even the proxies."

"We don't know enough," Margaret said. "I saw things in the bottom of the Rift, before I lost contact with the proxy. Big structures. And there's the anomalous temperature gradients, too. The seat of change must be down there, Arn. The parasite could be useful, if we can master it. The viruses that caused the immunodeficiency plagues are used for gene therapy now. Opie Kindred has been down there. He's suppressing what he has found."

"Yah, well, it does not much matter. They have completed synthesis of the metabolic inhibitor. I'm friendly with the organics chief. They diverted most of the refinery to it." Arn took out his slate. "He showed me how they have set it up. That is what they have been doing down in the Rift. Not exploring."

"Then we have to do something now."

"It is too late, Margaret."

"I want to call a meeting, Arn. I have a proposal."

Most of the science crews came. Opie Kindred's crew was a notable exception; Arn said that it gave him a bad feeling.

"They could be setting us up," he told Margaret.

"I know they're listening. That's good. I want it in the

open. If you're worried about getting hurt, you can always leave."

"I came because I wanted to. Like everyone else here. We're all scientists. We all want the truth known." Arn looked at her. He smiled. "You want more than that, I think."

"I fight my own fights." All around people were watching. Margaret added, "Let's get this thing started."

Arn called the meeting to order and gave a brief presentation about his research into survival of the exfoliations before throwing the matter open to the meeting. Nearly everyone had an opinion. Microphones hovered in the room, and at times three or four people were shouting at one another. Margaret let them work off their frustration. Some simply wanted to register a protest; a small but significant minority were worried about losing their bonuses or even all of their pay.

"Better that than our credibility," one of Orly Higgins's techs said. "That's what we live by. None of us will work again if we allow the *Ganapati* to become a plague ship."

Yells of approval, whistles.

Margaret waited until the noise had died down, then got to her feet. She was in the center of the horseshoe of seats, and everyone turned to watch, more than a hundred people. Their gaze fell upon her like sunlight; it strengthened her. A microphone floated down in front of her face.

"Arn has shown that contamination isn't an issue," Margaret said. "The issue is that the Star Chamber wants to destroy the reef because they want to exploit what they've found and stop anyone else using it. I'm against that, all the way. I'm not gengeneered. Micro-gravity is not my natural habitat. I have to take a dozen different drugs to prevent reabsorption of calcium from my bone, collapse of my circulatory system, fluid retention, all the bad stuff micro-gravity does to unedited Earth stock. I'm not allowed to have children here, because they would be as crippled as me. Despite that, my home is here. Like all of you, I would like to have the benefits of being a citizen, to live in the parklands and eat real food. But there aren't enough parklands for everyone because the citizens who own the habitats control production of fixed carbon. The vacuum organisms we have found could change that. The reef may be a source of plague, or it may be a source of unlimited organics. We don't know. What we do know is that the reef is unique and we haven't finished exploring it. If Star Chamber destroys it, we may never know what's out there."

Cheers at this. Several people rose to make points, but Margaret wouldn't give way. She wanted to finish.

"Opie Kindred has been running missions to the bottom of the Rift, but he hasn't been sharing what he's found there. Perhaps he no longer thinks that he's one of us. He'll trade his scientific reputation for citizenship," Margaret said, "but that isn't our way, is it?"

"NO!" the crowd roared.

And the White Mice invaded the room.

Sharp cracks, white smoke, screams. The White Mice had long flexible sticks weighted at one end. They went at the crowd like farmers threshing corn. Margaret was separated from Arn by a wedge of panicking people. Two techs got hold of her and steered her out of the room, down a corridor filling with smoke. Arn loomed out of it, clutching his slate to his chest.

"They're getting ready to set off the poison," he said as they ran in long loping strides.

"Then I'm going now," Margaret said.

Down a drop pole onto a corridor lined with shops. People were smashing windows. No one looked at them as they ran through the riot. They turned a corner, the sounds of shouts and breaking glass fading. Margaret was breathing hard. Her eyes were smarting, her nose running.

"They might kill you," Arn said. He grasped her arm. "I can't let you go, Margaret."

She shook herself free. Arn tried to grab her again. He was taller, but she was stronger. She stepped inside his reach and jumped up and popped him on the nose with the flat of her hand.

He sat down, blowing bubbles of blood from his nostrils, blinking up at her with surprised, tear-filled eyes.

She snatched up his slate. "I'm sorry, Arn," she said. "This is my only chance. I might not find anything, but I couldn't live with myself if I didn't try."

Margaret was five hundred kilometers out from the habitat when the radio beeped. "Ignore it," she told her pressure suit. She was sure that she knew who was trying to contact her, and she had nothing to say to him.

This far out, the Sun was merely the brightest star in the sky. Behind and above Margaret, the dim elongated crescent of the *Ganapati* hung before the sweep of the Milky Way. Ahead, below the little transit platform's motor, Enki was growing against a glittering starscape, a lumpy potato with a big notch at its widest point.

The little moonlet was rising over the notch, a swiftly moving fleck of light. For a moment, Margaret had the irrational fear that she would collide with it, but the transit platform's navigational display showed her that she would fall above and behind it. Falling past a moon! She couldn't help smiling at the thought.

"Priority override," her pressure suit said. Its voice was a reassuring contralto Margaret knew as well as her mother's.

"Ignore it," Margaret said again.

"Sorry, Maggie. You know I can't do that."

"Quite correct," another voice said.

Margaret identified him a moment before the suit helpfully printed his name across the helmet's visor. Dzu Sho.

"Turn back right now," Sho said. "We can take you out with the spectrographic laser if we have to."

"You wouldn't dare," she said.

"I do not believe anyone would mourn you," Sho said unctuously. "Leaving the *Ganapati* was an act of sedition, and we're entitled to defend ourselves."

Margaret laughed. It was just the kind of silly, sententious, self-important nonsense that Sho was fond of spouting.

"I am entirely serious," Sho said.

Enki had rotated to show that the notch was the beginning of a groove. The groove elongated as the worldlet rotated further. Tigris Rift. Its edges ramified in complex fractal branchings.

"I'm going where the proxies fell," Margaret said. "I'm still working for you."

"You sabotaged the proxies. That's why they couldn't fully penetrate the Rift."

"That's why I'm going—"

"Excuse me," the suit said, "but I register a small energy flux."

"Just a tickle from the ranging sight," Sho sid. "Turn back now, Dr. Wu."

"I intend to come back."

It was a struggle to stay calm. Margaret thought that Sho's threat was no more than empty air. The laser's AI would not allow it to be used against human targets, and she was certain that Sho couldn't override it. And even if he could, he wouldn't dare kill her in full view of the science crews. Sho was bluffing. He had to be.

The radio silence stretched. Then Sho said, "You're planning to commit a final act of sabotage. Don't think you can get away with it. I'm sending someone after you."

Margaret was dizzy with relief. Anyone chasing her

would be using the same kind of transit platform. She had at least thirty minutes' head start.

Another voice said, "Don't think this will make you a hero."

Opie Kindred. Of course. The man never could delegate. He was on the same trajectory, several hundred kilometers behind but gaining slowly.

"Tell me what you found," she said. "Then we can finish this race before it begins."

Opie Kindred switched off his radio.

"If you had not brought along all this gear," her suit grumbled, "we could outdistance him."

"I think we'll need it soon. We'll just have to be smarter than him."

Margaret studied the schematics of the poison spraying mechanism—it was beautifully simple, but vulnerable—while Tigris Rift swelled beneath her, a jumble of knife-edge chevron ridges. Enki was so small and the Rift so wide that the walls had fallen beneath the horizon. She was steering toward the Rift's center when the suit apologized and said that there was another priority override.

It was the *Ganapati*'s lawyer. She warned Margaret that this was being entered into sealed court records, and then formally revoked her contract and read a complaint about her seditious conduct.

"You're a contracted worker just like me," Margaret said. "We take orders, but we both have codes of professional ethics, too. For the record, that's why I'm here. The reef is a unique organism. I cannot allow it to be destroyed."

Dzu Sho came onto the channel and said, "Off the record, don't think about being picked up."

The lawyer switched channels. "He does not mean it," she said. "He would be in violation of the distress statutes." Pause. "Good luck, Dr. Wu."

Then there was only the carrier wave.

Margaret wished that this made her feel better. Plenty of contract workers who went against the direct orders of their employers had disappeared, or been killed in industrial accidents. The fire of the mass meeting had evaporated long before the suit had assembled itself around her, and now she felt colder and lonelier than ever.

She fell, the platform shuddering now and then as it adjusted its trim. Opie Kindred's platform was a bright spark moving sideways across the drifts of stars above. Directly below was a vast flow of nitrogen ice with a black river winding through it. The center of the Rift, a cleft two kilometers long and fifty kilometers deep. The reef.

She fell toward it.

She had left the radio channel open. Suddenly, Opie Kindred said, "Stop now and it will be over."

"Tell me what you know."

No answer.

She said, "You don't have to follow me, Opie. This is my risk. I don't ask you to share it."

"You won't take this away from me."

"Is citizenship really worth this, Opie?"

No reply.

The suit's proximity alarms began to ping and beep. She turned them off one by one, and told the suit to be quiet when it complained.

"I am only trying to help," it said. "You should reduce your velocity. The target is very narrow."

"I've been here before," Margaret said.

But only by proxy. The ice field rushed up at her. Its smooth flows humped over one another, pitted everywhere with tiny craters. She glimpsed black splashes where vacuum organisms had colonized a stress ridge. Then an edge flashed past; walls unraveled on either side.

She was in the reef.

The vacuum organisms were everywhere: flat plates jutting from the walls; vases and delicate fans and fret-works; huge blotches smooth as ice or dissected by cracks. In the light cast by the platform's lamps, they did not possess the vibrant primary colors of the proxy link, but were every shade of gray and black, streaked here and there with muddy reds. Complex fans ramified far back inside the milky nitrogen ice, following veins of carbonaceous compounds.

Far above, stars were framed by the edges of the cleft. One star was falling toward her: Opie Kindred. Margaret switched on the suit's radar, and immediately it began to ping. The suit shouted a warning, but before Margaret could look around, the pings Dopplered together.

Proxies.

They shot up toward her, tentacles writhing from the black, streamlined helmets of their mantles. Most of them missed, jagging erratically as they squirted bursts of hydrogen to kill their velocity. Two collided in a slow flurry of tentacles.

Margaret laughed. None of her crew would fight against her, and Sho was relying upon inexperienced operators.

The biggest proxy, three meters long, swooped past. The crystalline gleam of its sensor array reflected the lights of the platform. It decelerated, spun on its axis, and dove back toward her.

Margaret barely had time to pull out the weapon she had brought with her. It was a welding pistol, rigged on a long rod with a yoked wire around the trigger. She thrust it up like the torch of the Statue of Liberty just before the proxy struck her.

The suit's gauntlet, elbow joint, and shoulder piece stiffened under the heavy impact, saving Margaret from broken bones, but the collision knocked the transit platform sideways. It plunged through reef growths. Like glass, they had tremendous rigidity but very little lateral strength. Fans and lattices broke away, peppering Margaret and the proxy with shards. It was like falling through a series of chandeliers. Margaret couldn't close her fingers in the stiffened gauntlet. She stood tethered to the platform with her arm and the rod raised straight up and the black proxy wrapped around them. The proxy's tentacles lashed her visor with slow, purposeful slaps.

Margaret knew that it would take only a few moments before the tentacles' carbon-fiber proteins could unlink; then it would be able to reach the life support pack on her back.

She shouted at the suit, ordering it to relax the gauntlet's fingers. The proxy was contracting around her rigid arm as it stretched toward the life support pack. When the gauntlet went limp, pressure snapped her fingers closed. Her forefinger popped free of the knuckle. She yelled with pain. And the wire rigged to the welding pistol's trigger pulled taut.

Inside the proxy's mantle, a focused beam of electrons boiled off the pistol's filament. The pistol, designed to work only in high vacuum, began to arc almost immediately, but the electron beam had already heated the integument and muscle of the proxy to more than 400°C. Vapor expanded explosively. The proxy shot away, propelled by the gases of its own dissolution.

Opie was still gaining on Margaret. Gritting her teeth against the pain of her dislocated finger, she dumped the broken welding gear. It only slowly floated away above her, for it still had the same velocity as she did.

A proxy swirled in beside her with shocking suddenness. For a moment, she gazed into its faceted sensor array, and then dots of luminescence skittered across its smooth black mantle, forming letters.

Much luck, boss, SK.

Srin Kerenyi. Margaret waved with her good hand. The proxy scooted away, rising at a shallow angle toward Opie's descending star.

A few seconds later the cleft filled with the unmistak-
able flash of laser light.

The radar trace of Srin's proxy disappeared.

Shit. Opie Kindred was armed. If he got close enough
he could kill her.

Margaret risked a quick burn of the transit platform's
motor to increase her rate of fall. It roared at her back for
twenty seconds; when it cut it out, her pressure suit warned
her that she had insufficient fuel for full deceleration.

"I know what I'm doing," Margaret told it.

The complex forms of the reef dwindled past. Then
there were only huge patches of black staining the nitrogen
ice walls. Margaret passed her previous record depth, and
still she fell. It was like free fall; the negligible gravity of
Enki did not cause any appreciable acceleration.

Opie Kindred gained on her by increments.

In vacuum, the lights of the transit platform threw
abrupt pools of light onto the endlessly unraveling walls.
Slowly, the pools of light elongated into glowing tunnels
filled with sparkling motes. The exfoliations and gases and
organic molecules were growing denser. And, impossibly,
the temperature was *rising,* one degree with every five
hundred meters. Far below, between the narrowing perspective of the walls, structures were beginning to resolve
from the blackness.

The suit reminded her that she should begin the platform's deceleration burn. Margaret checked Opie's velocity and said she would wait.

"I have no desire to end as a crumpled tube filled with
strawberry jam," the suit said. It projected a countdown on
her visor and refused to switch it off.

Margaret kept one eye on Opie's velocity, the other on
the blur of reducing numbers. The numbers passed zero.
The suit screamed obscenities in her ears, but she waited a
beat more before firing the platform's motor.

The platform slammed into her boots. Sharp pain in her

ankles and knees. The suit stiffened as the harness dug into her shoulders and waist.

Opie Kindred's platform flashed past. He had waited until after she had decelerated before making his move. Margaret slapped the release buckle of the platform's harness and fired the piton gun into the nitrogen ice wall. It was enough to slow her so that she could catch hold of a crevice and swing up into it. Her dislocated finger hurt like hell.

The temperature was a stifling eighty-seven degrees above absolute zero. The atmospheric pressure was just registering—a mix of hydrogen and carbon monoxide and hydrogen sulphide. Barely enough in the whole of the bottom of the cleft to pack into a small box at the pressure of Earth's atmosphere at sea level, but the rate of production must be tremendous to compensate for loss into the colder vacuum above.

Margaret leaned out of the crevice. Below, it widened into a chimney between humped pressure flows of nitrogen ice sloping down to the floor of the cleft. The slopes and the floor were packed with a wild proliferation of growths. Not only the familiar vases and sheets and laces, but great branching structures like crystal trees, lumpy plates raised on stout stalks, tangles of black wire hundreds of meters across, clusters of frothy globes, and much more.

There was no sign of Opie Kindred, but tethered above the growths were the balloons of his spraying mechanism. Each was a dozen meters across, crinkled, flaccid. They were fifty degrees hotter than their surroundings, would have to grow hotter still before the metabolic inhibitor was completely volatilized inside them. When that happened, small explosive devices would puncture them, and the metabolic inhibitor would be sucked into the vacuum of the cleft like smoke up a chimney.

Margaret consulted the schematics and started to climb down the crevice, light as a dream, steering herself with the fingers of her left hand. The switching relays that controlled the balloons' heaters were manually controlled because of telemetry interference from the reef's vacuum smog and the broadband electromagnetic resonance. The crash shelter where they were located was about two kilometers away, a slab of orange-foamed plastic in the center of a desolation of abandoned equipment and broken and half-melted vacuum organism colonies.

The crevice widened. Margaret landed between drifts of what looked like giant soap bubbles that grew at its bottom.

And Opie Kindred's platform rose up between two of the half-inflated balloons.

Margaret dropped onto her belly behind a line of bubbles that grew along a smooth ridge of ice. She opened a radio channel. It was filled with a wash of static and a wailing modulation, but through the noise she heard Opie's voice faintly calling her name.

He was a hundred meters away and more or less at her level, turning in a slow circle. He couldn't locate her amidst the radio noise and the ambient temperature was higher than the skin of her pressure suit, so she had no infrared image.

She began to crawl along the smooth ridge. The walls of the bubbles were whitely opaque, but she should see shapes curled within them. Like embryos inside eggs.

"Everything is ready, Margaret," Opie Kindred's voice said in her helmet. "I'm going to find you, and then I'm going to sterilize this place. There are things here you know nothing about. Horribly dangerous things. Who are you working for? Tell me that and I'll let you live."

A thread of red light waved out from the platform and a chunk of nitrogen ice cracked off explosively. Margaret felt it through the tips of her gloves.

"I can cut my way through to you," Opie Kindred said, "wherever you are hiding."

Margaret watched the platform slowly revolve. Tried to guess if she could reach the shelter while he was looking the other way. All she had to do was bound down the ridge and cross a kilometer of bare, crinkled nitrogen ice without being fried by Opie's laser. Still crouching, she lifted onto the tips of her fingers and toes, like a sprinter on the block. He was turning, turning. She took three deep breaths to clear her head—and something crashed into the ice cliff high above! It spun out in a spray of shards, hit the slope below, and spun through toppling clusters of tall black chimneys. For a moment, Margaret was paralyzed with astonishment. Then she remembered the welding gear. It had finally caught up with her.

Opie Kindred's platform slewed around and a red thread waved across the face of the cliff. A slab of ice thundered outward. Margaret bounded away, taking giant leaps and trying to look behind her at the same time.

The slab spun on its axis, shedding huge shards, and smashed into the cluster of the bubbles where she had been crouching just moments before. The ice shook like a living thing under her feet and threw her head over heels.

She stopped herself by firing the piton gun into the ground. She was on her back, looking up at the top of the ridge, where bubbles vented a dense mix of gas and oily organics before bursting in an irregular cannonade. Hundreds of slim black shapes shot away. Some smashed into the walls of the cleft and stuck there, but many more vanished into its maw.

A chain reaction had started. Bubbles were bursting open up and down the length of the cleft.

A cluster popped under Opie Kindred's platform and he vanished in a roil of vapor. The crevice shook. Nitrogen ice boiled into a dense fog. A wind got up for a few minutes. Margaret clung to the piton until it was over.

Opie Kindred had drifted down less than a hundred meters away. The thing which had smashed the visor of his helmet was still lodged there. It was slim and black, with a hard, shiny exoskeleton. The broken bodies of others settled among smashed vacuum organism colonies, glistening like beetles in the light of Margaret's suit. They were like tiny, tentacle-less proxies, their swollen mantles cased in something like keratin. Some had split open, revealing ridged reaction chambers and complex matrices of black threads.

"Gametes," Margaret said, seized by a sudden wild intuition. "Little rocketships full of DNA."

The suit asked if she was all right.

She giggled. "The parasite turns everything into its own self. Even proxies!"

"I believe that I have located Dr. Kindred's platform," the suit said. "I suggest that you refrain from vigorous exercise, Maggie. Your oxygen supply is limited. What are you doing?"

She was heading toward the crash shelter. "I'm going to switch off the balloon heaters. They won't be needed."

After she shut down the heaters, Margaret lashed one of the dead creatures to the transit platform. She shot up between the walls of the cleft, and at last rose into the range of the relay transmitters. Her radio came alive, a dozen channels blinking for attention. Arn was on one, and she told him what had happened.

"Sho wanted to light out of here," Arn said, "but stronger heads prevailed. Come home, Margaret."

"Did you see them? Did you, Arn?"

"Some hit the *Ganapati*." He laughed. "Even the Star Chamber can't deny what happened."

Margaret rose up above the ice fields and continued to rise until the curve of the worldlet's horizon became visible, and then the walls of Tigris Rift. The *Ganapati* was a faint star bracketed between them. She called up deep radar, and saw, beyond the *Ganapati*'s strong signal, thousands of faint traces falling away into deep space.

A random scatter of genetic packages. How many would survive to strike new worldlets and give rise to new reefs?

Enough, she thought. The reef evolved in saltatory jumps. She had just witnessed its next revolution.

Given time, it would fill the Kuiper Belt.